*Power*Presentations

"Presentations that Sell, not Tell"

The Guide for Technology Sales Support

By

Patti Pace Fisher

ISBN: 1-4107-3336-X (e-book)
ISBN: 1-4107-3335-1 (Paperback)

This book is printed on acid free paper.

1stBooks – rev. 03/20/03

About the Author

Patti Pace Fisher has over 25 years of experience in the technology industry. She has served at both director and vice president level positions responsible for product development and product marketing. Much of her career has been spent in front of an audience giving product sales presentations and speaking to industry organizations. She has given over 2000 presentations with audiences ranging from a few to over 800 people.

She has been a guest speaker for such organizations as IBM, IHRIM, DPMA, Personnel & Training '87, IPM of South Africa, Grand Metropolitan Controllers Conference in the UK, East Tennessee Benefits Council, National Controllers Conference, Toastmasters of Georgia, and most of the large accounting firms. Patti has also conducted sessions at 100+ sales conferences and training sessions for her employers. She has worked for firms such as NCR, Management Science America, Energy Management Associates, ADP, Ceridian, and Computer Network Technology.

Patti originally wrote *Power*presentations in 1992 as a seminar guide for a product sales training module offered by her firm, P. M. Fisher & Associates.

Patti attended Northwestern State University in Natchitoches, Louisiana and she has an MBA from Emory University in Atlanta, Georgia. Over the years, Patti has been named as a member of two national awards, Outstanding Young Women in America and Who's Who Worldwide. Today, she is a technology-marketing consultant.

Introduction

This book provides a guide to presenting technology products and services to sophisticated buyers in a B2B environment. It does not and should not be used to replace product-knowledge training. It is imperative that any presenter thoroughly understand the product that he/she is selling. References will be made throughout this book regarding technical knowledge. If you are not confident with the levels of knowledge required, more product or marketing training will be a prerequisite to making these techniques work.

The techniques provided are suited for companies that use a team selling approach as well as those companies where the sales representative manages the sales cycle and presents the product. Team selling refers to the sales person being responsible for managing the account and closing the business while another person provides product and technical expertise throughout the sales cycle. There are obviously some differences when approaching a presentation and these will be identified throughout.

The primary objectives of this book are:
- Improve your ability to more effectively and efficiently sell your products.
- Improve your communication skills with your buyers.
- Improve your ability to control your audience.

- Improve your understanding of the dynamics of the selling situation.
- Manage the product sales cycle.

Contents

Chapter 1

The Basics of Public Speaking

This chapter discusses basic public speaking techniques. You should master these basic concepts before progressing to the selling techniques presented later. The topics covered are:

- Look the Part
- Organization
- Vocal Variety
- Body Language
- Eye Contract
- Grammar, Word Usage and Sentence Structure

Look the Part

It is important to remember that when you are in front of a buyer, you are an agent of your company. Your company has confidence in you and your abilities or they would not have put you into such an important role. As an agent, you must always look your best. Being neatly groomed is imperative. No shirts or blouses escaping your waistband, please! Men can wear tailored shirts that help solve this

1

problem. Women can tuck their blouses under the waistband of their pantyhose. Dress should generally be conservative. You want the audience to focus on what you are saying, not what you are wearing. This does not mean that you can't wear colorful clothes or accessories. Just make sure that they do not become a focal point for your listeners. Anything that will attract a buyer's attention should be avoided. The actual style of clothing may depend on your audience and what is acceptable dress for that particular company. During my many years of presenting, I usually wore a suit and my long hair up. But when I went into more relaxed companies, I wore a business dress and my hair down. The key is - - - you want your audience to be comfortable with how you look so they can concentrate on your presentation and not your looks.

Women have a unique set of challenges and as a result can alienate an audience with out realizing it. Makeup should be applied sparingly. Avoid wearing too much jewelry and do not wear jewelry that makes any noise (i.e. bracelets that jingle). False fingernails, if you wear them, should be reasonable in length. If you have long nails and you are fair-skinned, wear more neutral colored nail polish. Women with dark skin can wear dark colored polish without distracting the audience. It's also a good idea to avoid faddish clothes. More tailored looks give you a more professional appearance. The longer the time you will be in front of an audience should also dictate what colors to wear. Very bright colors such as orange and red can be stressful to an audience over time. Neutral tones and muted colors have a more relaxing effect.

Organization

If your presentation is to make sense to the audience, one it can follow all the way to a conclusion, it must be organized in a logical form. This logical form should be one that the audience recognizes quickly and that relates to their business. Presenting your product the way it is technically structured, for example, will lose your audience. Organization is really nothing more than clearly putting your ideas together in an orderly manner with a focus on your audience.

As your company's presenter, your business is selling ideas and as anyone in sales will tell you, success comes only when you carefully plan and organize your approach. You must clearly identify the key problem for the audience and then lead them logically toward an acceptable solution to that problem. Merely talking around the subject in a haphazard manner will leave your listeners confused, not convinced.

As a seller of ideas, you must always speak from your audience's point of view. They will be motivated only by what they want, not by what you want. As you organize your talk, think in these terms. Analyze what it will take to motivate your audience, to get them to agree with you and to understand you or take action on your behalf. Then, develop your ideas to best supply that motivation.

Your presentation should have a clearly defined opening, body and close. I have seen so many presentations that do the first two, but leave out the close. This can be disastrous! That's your time to ask for action.

The opening of your presentation should be designed to catch your audience's attention. It should also lay the groundwork for the remainder of your presentation. It is important for the audience to know why you are there, what you are going to tell them and why they should listen. It should also provide a roadmap for your presentation, reducing the opportunity for the audience to interrupt the flow of your talk.

You should avoid these common weaknesses in your opening.
- An apologetic statement.
- A story that does not relate to your topic.
- A commonplace observation delivered in a commonplace manner.
- A long slow moving statement.
- A trite question such as, "Did you ever stop to think?"

- A boast about your company or product (you haven't earned the right yet).

The body of your presentation should build on your opening and provide factual support for it. The body of your presentation will be covered in detail later in this book. Some general tips to remember:

- Be a storyteller.
- Relate examples to the buyer's environment.
- Talk in a conversational mode.
- Do not read your visual aids.
- As your talk progresses, tie in topics mentioned earlier to reinforce your points.
- When you describe a feature, follow it with a benefit.
- As you go from one general area to another, briefly summarize before you introduce the new topic.
- Show enthusiasm.
- Show genuine interest in their needs and problems.
- Try to make each person in the group feel important.

The close of your presentation should summarize the major benefits of your solution as related to the buyer's needs. It should be brief and hard-hitting. Don't waste your time on what you think they should remember, but spend your time on what is important to the buyer. Specific recommendations will come later.

5

Vocal Variety

A good speaking voice should be a balance between different levels of volume, pitch and rate while having a pleasing sound quality. Let's examine each of these individually.

<u>Volume</u>

Some people have an unconscious habit of always speaking too loudly. Such people should make a conscious effort to speak more quietly. Loud voices make people feel uncomfortable and that they are being talked at, rather than being talked with. Plus, it's annoying. On the other extreme, there are people who can barely be heard. They need to work on projecting their voices. You don't want your audience to have to strain to hear you because they probably won't. They will just tune you out all together. People who speak too softly also project an air of insecurity. You want your audience to believe that you are confident in yourself and your product. So, don't overwhelm them, but make sure they can hear you.

You should also avoid predictable changes in volume. People often speak louder at the first of a sentence and then lower the volume toward the end of a sentence. Or, the reverse can be true. This gives your speech pattern a singsong effect, which detracts from what you are saying.

Changes in volume can be used very effectively to add variety and emphasis to your presentation. Speaking at the same volume continuously will become monotonous.

Remember, it's the content of your presentation you want your audience to remember not your annoying habits. You also want the audience to remember you as someone they enjoyed listening to.

Pitch

Good speakers vary the pitch of their voice to convey emotion and conviction. Too high a pitch should always be avoided, because it suggests immaturity and excitability. The best approach is to make a conscious effort to be conversational in your speaking.

Rate

The most effective speaking rate falls into the range of 125-160 words per minute. You can easily keep within this range by speaking rapidly enough to avoid a boring drone, yet slowly enough to be clearly understandable. Vary your speaking rate during your presentation to reflect new ideas and to emphasize points of the presentation. You should slow your rate when introducing new terminology that may be unfamiliar to your audience.

Quality

The most important recommendation for voice quality is to relax your throat while you speak. Think in terms of friendliness, confidence and a desire to communicate. Relax and remove any tension from your voice. A pleasing quality of tone will usually follow. A good speaking voice generally has the following characteristics:

- The tone is pleasant, conveying a sense of friendliness.
- It is natural, reflecting the personality and sincerity of the speaker.
- It has vitality, giving the impression of force, confidence and strength, even when it isn't especially loud.
- It portrays various shades of meaning, never sounding monotonous or emotionless.
- It is easily heard, due to both proper volume and clear articulation.

Body Language

A most effective way to communicate your sincerity is to put your whole body into the presentation. An audience will usually believe what they see in your face and mannerisms in preference to what you tell them in words. Therefore, you need to develop skill in the use of body language to insure that the audience receives the same message through their eyes and through their ears.

The body language that you use to enhance your presentation should include facial expressions and body movements to show enthusiasm or boredom, pleasure or pain, sincerity or sarcasm. Plan to use both facial expressions and body movements in your presentations. The most expressive part of body language, when you are speaking before an audience, is the use of your hands and arms to illustrate your words. In many situations, these are the only visual expressions that the whole audience can see.

Some basic gestures show:

- Size, weight, shape, direction, location
  These physical characteristics call for hand gestures.

- Importance or urgency
  Show your audience how vital your point is. For example, hit your fist into your open hand or gesture to punctuate your points.
- Comparison in contrast
  Move both hands in unison to show similarities. Move both hands in opposite directions to show differences.

Things to remember:

- Each gesture should be large enough and definitive enough to be seen by everyone in the audience, but not so exaggerated that it distracts from your talk. A gesture can be considered good if it helps the audience understand your message, bad if it draws attention only to itself.

- Avoid using the same gesture over and over.
- Avoid pacing up and down in front of the audience.
- Avoid planting yourself in one spot throughout the presentation. Do not "anchor" yourself to a podium or table.
- Avoid turning your back on the audience. If you need to write on a whiteboard that requires turning your back on your audience, write first then turn facing the audience to speak.
- Move toward the subject (person) of a question.
- Avoid nervous movements of the hand or playing with a pointer or pen.
- Don't hold a pointer, pen or other device unless you are using it.

When you include body language as part of your presentation, you are adding another

dimension to it. You are showing the audience what you mean as well as telling them.

Your talk comes across in a relaxed, natural manner, as if you were having a conversation

with your audience. By giving these techniques a try, you can begin to understand the principles of timing, accuracy and proper magnitude in your body's gestures.

Eye Contact

In addition to the use of body language, eye contact is very important if you are going to include the entire audience.

Good eye contact requires that you:

- Look directly at each individual during the presentation.
- Avoid looking over the audience's heads.
- Try not to direct the presentation to one or two people. Move your eyes from section to section looking at each individual in the room.
- During audience questions, look directly at the speaker.
- Avoid being distracted by continual motion, someone tapping their pen or someone swinging their leg.
- Recognize new people as they enter the room.
- Use your eye contact in conjunction with your body language.

Correct Grammar, Word Usage and Sentence Structure

Your primary objective during your presentation includes: 1) selling yourself, 2) selling your company, and 3) selling your product or service. In order to sell yourself, you must meet certain criteria in the eyes of your audience:

- Professional
- Personable
- Articulate
- Intelligent
- Knowledgeable
- Believable

Incorrect grammar, word usage and sentence structure can undermine your credibility. To some individuals, it may even be considered offensive. By organizing and practicing your presentation, you will avoid the common pitfalls.

Some examples are:

- Use of "ah" and "um".
- Connective "ands".
- Awkward sentence structure.
- Inappropriate word usage or the use of non-existent words such as "irregardless".
- Over-use of particular phrases or words.
- Poor grammar.

Summary

You will only get one opportunity to make a first impression. Therefore, looking professional is key. If you look the part, you have opened people up to listening to you. Good organization will keep you and your audience focused. Body language, vocal variety and eye contract improve the audience's receptiveness to your presentation. And finally, correct grammar, word usage and sentence structure add to your credibility and greatly improve the chances of the audience understanding your message.

These techniques will work whether you are presenting your product or service to a prospective buyer, selling an idea within your company or delivering a speech to a trade organization. In other words, these techniques will work any time you need to deliver a message to one or more persons. It is the combination of techniques that make this approach work. Implementing only one or two of these will result in a less than effective presentation. Remember that your audience is going to evaluate your performance based upon their overall impression of you, not just one aspect of your presentation.

Chapter 2

Surveys

There are several common obstacles to successfully selling technology products and services. This and the next few chapters will discuss these obstacles and how you can overcome them.

The first set of obstacles is primarily due to poorly planned preparation. Specifically they are:

- Unprofessional speaker.
- Making assumptions about the buyer's needs.
- Not understanding the buyer's value chain.
- Trying to "tell all" about the product or service.
- Lack of knowledge of the evaluation team dynamics.

Chapter 1 discussed in some detail how to improve your professional speaking skills, which will help you overcome the first obstacle list above. This chapter is designed to ensure that you do not make the other four mistakes that will undermine even the most professional, personable speaker.

The survey process will enable you to learn the specific needs of your buyer and provide you with the information that you need to tailor your presentation. This will keep you from trying to "tell all" about your product. By understanding the buyer's value chain, you can align your company and your product to be in sync with your buyer's competitive position. And finally, during the survey process you can gain valuable insights to the dynamics of the evaluation team members. Each of these topics will be discussed in detail.

Understanding the survey process.

A survey of the buyer project team prior to making your presentation will enable you to present your product or service in its best possible light and to ensure that it is presented relative to the buyer's needs and expectations. Your goal is to make the buyer want and need your product. I use "want" first because, if the buyer doesn't want your product whether or not they need your product is irrelevant. In the case of new technology or a brand new idea, you may have to create a need that will generate a "want" through a soft education process. This is most easily done when you are dealing with an existing customer that is happy with your company's product or service they already use. Educational sessions work best when they are promoted as such. User meetings and seminars are good forums.

Both the sales and sales support person in a team-selling environment can do the survey process. However, they should be separate events and with different buyers. Buyers usually have a pre-existing bias towards a sales representative. They know that the sales representative is going to try to sell them something. Fancy titles for sales representatives are not going to overcome this bias. In the situation where the sales representative also presents the product, he/she must be sensitive to the buyer's perception during the survey and the presentation.

Surveys should be done well in advance of the presentation to ensure that you have adequate time to prepare. Nothing will destroy your credibility faster than to do a survey and present your standard presentation without using what you learned. If you do this, your are telling the buyer that 1) his/her time was not of value; 2) you are not particularly interested in the buyer's business problems; and 3) you are not what you earlier presented yourself to be. This will cause the buyer to resent the pre-presentation time they spent with you. If you don't place value on your buyer's time, why should you expect for your buyer to place value on your time. Every presentation you do in a selling situation should be tailored to the audience. This even applies to seminars with mixed audiences, but this will be addressed in Chapter 9, Seminars.

The objective of the survey is to prepare you for the presentation. This should be your ONLY objective and you must keep this in mind while you're doing the survey. Don't let the buyer get you side-tracked. I often hear the complaint that the buyer doesn't want to spend the time to do a survey. This is a symptom of one of two things: 1) you haven't presented it properly or 2) you don't have a qualified buyer. If this presentation is not a priority for them, don't waste your time.

If you are getting resistance, you may want to try the following:

> *I am committing my company's resources to present our product to your company. Before making this commitment, I want to be sure that our solution is the best one for your particular needs. Every company is unique and in order for me to do the best job for you and for my company, the survey is a prerequisite. I do not want to waste your time or mine. I am not asking for a large block of time. A few minutes with each of the project team members and the decision makers will enable me to be prepared and present our company's solution in a professional, knowledgeable manner.*

By using this approach, you are showing the buyer that you are a good business person and that you value your company's resources. If the buyer is really a prospect and not a suspect, they will agree to the survey. If they don't, they are probably kicking tires; have selected another vendor and want to use your product as matrix fodder; or are not committed to this project. If they are not willing to let you do your very best, don't waste your time. It will only reflect poorly on you in the long run.

Surveys are best done on-site in the buyer's environment. If you can meet with each team member in his/her own office, he/she will feel more comfortable and will probably open up more to your questions. By seeing where people work, you will also get a good feel for the pecking order of the project team members. It will also give you an opportunity to start building rapport for when it will work to your advantage, the presentation. You should limit your time with each person, respecting the original ground rules, 15 minutes for example. This means you need to be prepared.

When setting up a schedule for the meetings, make sure that you allow time to get from one interviewee to the next and remain on schedule. Otherwise, you will create a ripple effect on your schedule and be late to appointments. You will need a point person at the buyer's company to reschedule meetings that buyer team members don't make. Preferably, the point person can reschedule them around your schedule and on the same day. If buyer team members within a category are not meeting with you, you have a problem and should reconsider this buyer.

When economics or timing rule out an on-site visit, telephone surveys can also be effective. The same rules apply. Be prepared and stick to the schedule.

The survey should include the following key elements at a minimum:

- Does your solution meet their needs? How much of the product (features) do they need? Do they want?
- Who is interested in what and why?
- Is there any pride of authorship with existing solutions?
- Who are the decision makers? Who are the influencers? Do they have the same agenda?
- By individual, what is their primary area of interest?
- What are their internal "buzz" words, terminology?
- How do members of the team interrelate? Reporting relationships? Personalities?
- Who is your major competitor? Who else is the buyer evaluating? Any biases toward a particular vendor?
- Where are the land mines?
- Can I find a friend?

These key elements are not listed in any specific order or priority. You goal should be to answer all of them during the course of the survey. In an ideal solution, this will happen assuming that you are talking with the right people. Other times, you will have to glean this information as the evaluation progresses. In any event, the order will depend on the sequence with which you talk with people and their willingness to directly or indirectly tell you what you need to know. This list assumes that you have reached this stage knowing that the buyer has some need and more importantly has the money allocated to buy your installed product. The above list is not inclusive as there will be additional information that is specific to the type of product that you sell. Let's discuss these key elements and the "whys and hows" of using this information.

In a team-selling environment, the sales support/product person should be responsible for ALL product related issues.

The buyer evaluation team will be comprised of several levels of people. The decision-makers and the influencers are, of course, the most important. However, it's vital that you get input from the worker level people as well. Keep in mind, a worker level person may not be able to close the deal for you, but they can certainly blow it for you.

Does your solution meet their needs? Their wants?
One obvious reason for this question is does your product fit their needs? Should you continue to pursue this buyer? Rarely does a product fit 100% of a buyer's needs out of the box. In addition, you are not dealing with a buyer's actual needs necessarily, but with their

perceived needs. Therefore, within your company reasonable-fit guidelines need to be established based upon your ability to modify the product at a cost that is not prohibitive to the buyer or your company's profit margins. Your marketing department should provide you with these guidelines.

But what about their wants?

This is often a difficult issue to work with, especially if the buyer's wants are unrealistic or cost prohibitive. If you can eliminate these during the survey, you will be doing yourself a tremendous service. A question you might ask: "If the product could do your 'want', how would that help you in your job?" Try to get the buyer to qualify or quantify this 'want'. Often data capture 'wants' are too costly to justify in terms of value. You might ask, "If you had that data, what kind of reports would you produce? How often would you require them? Do you need them to run your business?" What you need to try to do is eliminate the frivolous features and get them to focus on what they need to be successful and within their implementation timeframe.

Unless they are really headstrong or slightly "brain dead", there is a good chance you won't hear this 'want' during the presentation. If you asked the tough questions during the survey, they won't want to be put on the spot in front of the group during the presentation. Your attitude while asking these questions is very important. You have to position yourself as the uninformed one and you are asking the buyer to educate you on the importance of the 'wants'. If you play your cards right, you will let them talk themselves out of the frivolous features and help them focus on the real needs. The real needs are what you need to know for your presentation.

How many of your product features/functions do they need? Do they want?

As vendors we know a tremendous amount of detail about our product and services. We are proud of our product. And, we can't wait to tell people about it! This attitude is a doubled-edged sword. The more you know about your product the better. If you are going to be an enthusiastic presenter, you have to be proud of your product and eager to spread the good news.

What should be a positive attitude can turn into a negative if the quantity of product knowledge presented is not controlled. The natural tendency is to tell everything you know about the product because it a wonder product and YOU are an expert! And, you want everyone to know it!! You have to learn to control this urge if you want the buyer to respond to you in the way that you intend.

Too much information is worse than not enough. Listeners can only absorb so much information during the course of your presentation. Therefore, you need to ration all of your product knowledge down to the buyer's need to know. They are going to tell you what they need to know during the survey. That's what you should focus on during your presentation.

If you missed something during your survey and it's important to the buyer, they will ask. Then you can expand your talk to include this issue.

Who is interested in what and why?

During your survey, you need to identify specific needs for each person that you interview. These need to be interpreted by you into a feature of your product with a benefit for the buyer, the company and the individual that had the need. Then during your presentation when you present the feature you present it as a need by an individual that you name and you can follow it with a benefit for that individual or the company. This involves your audience, makes the person who had the need feel important and heard, plus it relates your product specifically to the buyer's needs. This can go a long way in making the audience relate to your product and begin to think of it as their solution. It also reinforces that you understand their business needs and that their needs are important to you.

Is there any pride of authorship with the existing solution?

Wow, talk about a land mine! If you don't know about this, it can literally blow up in your face during a presentation. If there is pride,

23

it will be important to acknowledge the author during your presentation and make some positive comments about the existing product or solution. You want to avoid being patronizing. You can comment on unique

features specific to their environment, a well thought out design that has survived time, or special reports for example. It's important that what you praise is worthy of it. You need to be and sound sincere. If you are not, you can depend on the audience knowing that you're not. When you find the author during the survey, be sure to get some examples to use. Just let the author talk. He/she will be sure to brag a little OR a lot. Take notes and be interested.

Who are the decision makers? Do they have the same agenda?
You should ask everyone you talk with this question, "Who will be making the final decision?" and follow it by, "What is his/her most important criterion for making this decision?" If the answer to this first question is, "the project team will decide." Or "I will." And if he/she is not an officer of the company, you have not gotten the correct answer. Don't press the issue. You will find out during the process who the influencers are and who the decision makers are. It's important to prioritize the needs of these two groups first during your presentation. Try to include everyone else, but don't forget who is going to get you the business.

You may also find out that there may be more than one decision maker or officer that can make or break the deal. It's important to know what each one considers his/her major objective with this project. They may not be the same. You had better be sure that you know what each one's agenda is and how your product can fit that agenda. Depending on the magnitude of the impact of your product on an organization, this decision could make or break careers in the buyer's organization.

By individual, what is their primary area of interest?

It is important to know by individual what their primary area of interest is in the solution that you are proposing. It may be outside their area of responsibility, but it may have a direct bearing on their ability to do their job. Make sure you find out who benefits from the solution and why.

If you are under a time constraint for your presentation, it's a good idea to develop a short list of attendees and their areas of expertise and interest. Then as you present your product, you can identify individuals in the audience as beneficiaries of particular features. This is also helpful if the group is large, and your focus is on the influencers and decision makers, you can still involve non-essential personnel in the presentation.

What are their internal "buzz words" or specific terminology?

25

Every industry-based conversation is loaded with terminology unique to that industry. The same is true within a company. It's important for you to know the buyer's internal terminology as it relates to your product. If you talk about your product in terms of your company's internal terminology, even if you explain, you are asking the buyer to interpret your words every time you use them. They won't do it. They'll just tune you out. Use their terms and their "buzz" words in place of yours. They have enough to deal with learning about your product without having to learn the product lingo.

Using their terminology is another way for the buyer to begin to think of your product as their solution.

How do members of the team interrelate? Reporting Relationships? Personalities? Understanding the dynamics of the group will help you considerably during the presentation. How the team interrelates will influence your presentation room setup and seating. This will be discussed in detail in Chapter 8, Controlling and Analyzing your Audience. Knowing the personalities prior to the presentation will also help you to control the presentation. For example, long winded individuals should be asked closed ended questions. People with a sense of humor can come in handy if the group gets rather "deadly" or cynical.

It's also important to know where people report organizationally. It will allow you to prioritize their importance to the process and the time you allow them to participate. You must be polite and empathetic to all participants, but your energy should be focused on the influencers and decision makers. Don't let worker-bees dominate your presentation. They may have a tendency to get bogged down in unimportant details or details that would be better addressed outside the main presentation.

Who is your primary competitor with this buyer? Who else are they evaluating?

In many situations, the buyer will be reluctant to discuss this with you. However, it never hurts to ask. If they won't tell you, back off. Whether or not they tell you directly, you can usually find out by asking some pointed questions that, without naming a competitor, sniff them out. The buyer may also use another vendor's terminology without realizing it. You need to be alert to such slips. The more you know about your competition, whether it's another vendor or another alternative, the better off you are. What you are looking for is an opportunity to insert your competitive advantages or disadvantages during a presentation in a positive manner and without using the competitor's name. However, if you have many competitors, you need to know which ones you are competing with in this buyer situation. Otherwise you bog down your presentation, dilute your product's value and confuse your audience. You don't want the audience thinking, "Why is he/she talking about that?"

Where are the land mines?

Land mines are everywhere. They can be planted by competition, by someone on the project team or by you.

If you are knowledgeable about your competitors, you should know in advance what land mines they are going to plant, just as they will know which ones you are going to plant. Your marketing department should be tracking your competitors and should have developed strategies for each known land mine. Be sure that you know these strategies and they are credible to you. If you don't believe the strategies, neither will your audience. The best way to address competitive land mines is to weave your competitive strategy into your discussion of a feature or function. For example, "After lengthy discussion with our customers, we chose this approach because . . ."

Buyer land mines are tougher to detect. Sometimes they come out during the survey process. If this is the case, you can plan strategies to handle or neutralize them during the presentation. If it's a large enough mine, you may decide to walk away rather than spend more time and money on a lost cause.

If a major land mine explodes during the presentation, you have two options 1) think quick on your feet and diffuse it or 2) call for a short break in the presentation. If the short break is your only alternative, meet with the decision makers in a casual setting to determine if it is in everyone's best interest to stop or if it is worthwhile to continue. The decision makers will appreciate your consideration of their time and your business-like approach. You should only continue if it is in YOUR best interest. If stopping the meeting is appropriate, reconvene the group and in a polite, professional manner announce to the group that after a discussion with the decision makers (by name), we have decided to end the presentation. If a future meeting has been or will be scheduled when the issue is resolved, this should be announced as well. Thank everyone for their time and effort. It's best at this time to make a clean getaway. Side conversations with other members of the project team are not a good idea. A telephone conversation at a later time may be appropriate. The very worst land mines are the ones you plant yourself and then rather ungracefully step on. This can easily happen to an inexperienced presenter. If you are new at this, I recommend that you role play in front of experience product presenters that can set you up in a non-hostile, non-costly situation. Usually only practice helps eliminate this unfortunate event.

During the survey, make a note of all potential land mines that you might plant. If the survey reveals a feature or function that your product does not do particularly well, you will want to plan for this in your presentation. It usually means that you want to de-emphasize this specific feature while emphasizing another. This helps you avoid detailed questions about something that you don't want to discuss. If you do get questioned, answer with a "yes" or close ended answer and move on. Do not leave the door open for more questions or for a more detailed answer.

Many times presenters set land mines by selecting a poor choice of words. "Fixing bugs" for example, is a negative form of saying "We continually enhance our products" when you are discussing new releases of your product. Even though we know that 100% perfect products are rarely if ever delivered and "bugs" is a common term, you should avoid using any words that may bring doubts to your buyer's mind. Any phrases or terms that may be misconstrued should also be avoided. Think in positive terms and use positive words.

Can I find a friend?

Finally, you should try to find a friend among those that you interview. This should be either an influencer or a decision maker. No one else will do. This should be a person that has a positive attitude, speaks with authority and appears supportive during you interview process. It will generally be someone who is willing to volunteer information, especially about the process of the evaluation rather than the product. What you are looking for is someone who will be supportive of YOU during the presentation and help you out when you don't understand a question or will reinforce what you are saying during the presentation. If you think you have someone who has potential, you should ask for their support. An unassuming way to solicit support is to approach the person something like this. "This presentation is very important to me. I would appreciate your advice. Is there anything that I should know before making this presentation that would help me be more effective?" Most of the time, you will get a positive response. If you are given advice, you must be sure to use it. In the best case, you'll find out if your competition has done something that affected your buyer in a negative way. You want to make sure that you don't make the same mistake.

Developing the Components of Your Survey

Once you have your survey developed, you should be able to use it for every account. You will find that over time you will need to revise the survey as you become more skilled at conducting one and as your business climate evolves. Now that you understand how you will use the survey, the next step is planning the survey.

There are two major sections to your survey: 1) the buyer's organization and 2) the buyer's product needs.

The Buyer's Organization

In order to effectively present your product, you must first understand the buyer's value-chain. The value-chain relates directly to how your buyer develops strategies or reacts from a competitive standpoint with specific activities. A specific activity will create different economics, have a high potential impact of differential or represent a significant or growing proportion of cost. This concept is explained in detail in Michael Porter's book <u>Competitive Advantage</u>. What this means to you is that you must relate your product to how it impacts your buyer's competitive position. This will vary by industry. To be competitive your buyer must achieve cost leadership or product/service differentiation or both. If you understand the buyer's business and what is important to the company, you can relate your product to their value-chain. As a result, your product appears in sync with their strategy and they will be able to see and relate to the benefit of choosing it.

In order to do this, you must understand their business. In a team selling environment, this should be the primary responsibility of the sales representative. You should know what product/services they produce and how they consider themselves positioned in the market. For example, are they the Wal-Mart of their business or the Neiman Marcus? In this example, you would know whether price or image is the most important. When you know this, you know how to position your company and product. You want to align your company with their value-chain. Their annual report is a good starting place prior to actually visiting the company for a survey. And depending on the type of business, company product/service brochures or a review of their website are good sources of information. These should give you a good flavor for the buyer and will demonstrate their value-chain if your conversations have not readily identified what their company strategy is.

You should also understand how the company is organized, who the officers are, which ones are impacted by the choice of your product and which ones will be involved in the decision making process.

The company's organization may have a direct effect on how the evaluation will evolve and also whether it creates challenges for your product or service. Centralized versus decentralized, domestic or global, diverse multi-company situation and other company organizational structures may impact the complexity of your evaluation, the final decision making and your product fit. Early on and preferably pre-survey these issues need to be understood and accepted by you as manageable.

Have your primary contact draw you an organization chart. It should focus on the organizations directly impacted by this evaluation and those directly above and below. Once done, walk the buyer back through the diagram and at this time you can ask your questions about the business and about the people. Unless personalities, rivalries and conflicts are volunteered, don't ask. These questions are better left to the individual interviews. While learning about their organization, you will want the project team defined in terms of who they are, their titles, their job responsibilities, their project responsibilities and any constraints on your access to them. Be leery of situations where your access is severely limited.

If someone is impacted by this decision and is not involved in the decision, this should send up red flags. You have a potential land mine. This needs to be discussed with the officer who will be signing the contract. If your contact does not want this person directly involved, you might suggest that since he/she will be impacted by the decision that you send the uninvolved officer copies of your correspondence just to keep him/her informed. If you are still getting resistance, you have a problem. There are multiple agendas going on here (politics) or this project does not have the right support within the organization. Is it a real deal?

The Buyer's Product/Service Needs

In a team selling environment, this activity should be done by the product person and preferably without the sales person present. Product people are not considered sales people, even though we know they are. They understand the product and can relate better to a day-to-day technically involved person. They have credibility without posing a threat. They won't ask for the business. They are asking for help and people like to help. People like to feel important and people like to tell you what they know.

You should have a pre-printed set of questions. It is best to have several questionnaires broken down into areas of responsibility based upon how the buyer does business. These should not be too detailed. You are not looking for their technical specifications. That should come later. Remember you are there for only one reason. You should include mostly open-ended questions. The more they talk and the less you do, the better you will come across. Keep in mind that you are NOT selling the product in this meeting. You are gathering information. If you are asked product questions, answer "yes" or "no", but do not explain. If you are pressed to explain, politely answer that this will be covered in detail in your presentation, but that you are on a tight schedule today and move on to your next question.

While they are answering your questions, you will need to probe to find out what is really important to them. Some example questions are:

- Is that important to you?
- How will that affect your job?
- Does this also affect Mr. Johnson's (real name) job?
- How do you do that now?

- If you could do that, how much time (or any quantitative measure) would that save you?

During your interview, you should try to get examples of input and output from their current or proposed solution if it is relevant to your application. This will enable you to customize your presentation materials and talk with data that is already familiar to your buyer.

You can tie in competitive questions if they mention a feature that you know is unique to particular vendor by simply saying, "Oh, you must be looking at XYZ". All they can say is "yes" or "no". You need to listen for competitors' terminology. If you hear it, you don't need to pose the question. If you're very comfortable with the interviewee, you can ask them directly who else they are looking at, but wait until the end of the interview. You don't want them to clam up.

Keep in mind that you should be relaxed and conversation during this interview. You do not want to come across as someone conducting an inquisition.

Summary

I cannot stress enough how important the survey process can be in helping you give effective product/service presentations, presentations with power. If you go in unprepared, you cannot be your best. In today's competitive environment, you will be dealing with sophisticated buyers and aggressive competitors. Understanding your buyer's value-chain, people and business needs and using that data in your presentation will greatly enhance your ability to close business.

Special Note: There will be times when you find yourself in the dreaded RFP situation and have limited if any access to the evaluation team. If a buyer has developed an RFP, there is a high probability that a vendor has already been chosen. Unless you were the vendor helping the buyer with the RFP, as a general rule, you should be restrained about devoting substantial resources to this buyer.

Chapter 3

Pre-Planning Your Presentation

The organization of your presentation will be critical to your success. There are several factors that should be considered before actually preparing the visuals and the physical characteristics of your presentation. They are:

- Timing of presentation within the entire sales cycle.
- Time allocated for the presentation.
- Time constraints on key buyers.
- Physical location.
- Size of audience.
- Timing of your presentation relative to your competitors.
- Availability and appropriateness of your company's resources.

Let's examine these factors and how they will impact and influence how you prepare for your presentation.

The timing of your presentation within to the sales cycle

Depending on the type of product that you are selling, the sales cycle will vary in length. In most cases you will have some history to indicate the average timeframe it takes from initial contact to the signing of the contract. This too may vary depending on the size and complexity of the buyer's situation. Your marketing department should be tracking sales cycle activity and depending on where your buyer is relative to their purchasing cycle, it will impact how you approach your initial and possible future presentations.

A common mistake is to do an extremely detailed presentation too early in the sales cycle. First, the buyer is probably not ready for detailed information and secondly, you won't have a strong reason for going back into the account. It is the responsibility of the sales representative to determine early in the sales cycle how the buyer plans to evaluate the product. Distinct stages need to be defined and understood by both parties. If the sales cycle is going to extend over several months, you need to define internally how you want to manage this buyer over this period of time.

Most sales training courses will teach you that you must sell in the following sequence: 1) sell yourself, 2) sell your company, and 3) sell your product or service. If you do these out of sequence, you will get yourself into trouble. This is valid even with well known companies, because the buyer needs to understand the specifics of a future relationship between their company and yours. You need time to establish a rapport with and an understanding of the buyer. You can not do this during a detailed presentation. People will buy from people that they like and make them feel comfortable and whose product they feel confident will meet their needs.

If you know that you are dealing with an extended evaluation, your first presentation should focus on selling yourself and your company while introducing your product. Therefore, the product portion of the presentation should be high level and buyer value-chain dependent. The same processes apply in terms of organizing your visual presentation; it will just be on a higher level.

The time allocated by the buyer for your presentation
You may not always be afforded the amount of time that you would prefer to do your presentation. Being flexible is key. I have seen people in this situation that obviously are used to having 3 hours for example to present and when the time is cut back to 2 hours, they merely squeeze 3 hours worth of information into the 2 hours by talking faster and skipping much of the prepared material. I do not recommend this approach. Your presentation will appear rushed and the audience will feel your tension. Plus, there will be little time for interaction. This is not good for you or for your audience.

Only prepare the amount of material that you can conversationally cover in the allotted timeframe. This usually means that you will cover less material and therefore, it needs to be more specific to the buyer's needs. Features/functions that are nice to have will have to wait for another presentation or printed materials. Be sure to allot enough time for open discussion and audience involvement. No ones likes to come to a presentation and get a lecture. It's boring and it's condescending.

Time constraints on key buyers

This is a question that is often overlooked. Don't assume that everyone who plans to attend will be present during the entire presentation. People are busy and this product evaluation, although important, may not be a company officer's #1 priority. Find out in advance if someone important does have a time constraint. Verify this with your audience before you start your presentation. Then, make sure you cover what this person needs to hear before he/she has to leave.

Physical location

If you are presenting at the buyer's location, ask to see the meeting room before presentation day. You need to know the size and shape of the room and the available equipment if you should need some. Get the name of an administrative person that can help you with meeting room needs the day of the presentation. On presentation day, get there early so that you can setup well in advance of the presentation and work out any equipment problems should there be any. Room setup will be covered in Chapter 6, Physical and Professional Environment.

If you are meeting off-site, not in your company offices, the advice above applies. In your own office you have complete control and should be well prepared.

Size of the audience

The size of your audience will impact how you set the room up and your visual aids. These will be discussed later in Chapter 5 and Chapter 6. The size of your audience will also dictate the level of detail that you can present. It is very difficult to do a detailed technical presentation to a large group. This is due primarily to the multiple levels of expertise that will be in the audience. You will have to find a middle ground or risk losing most of your audience during your presentation. It may be more effective to segment your presentation in order to address multiple knowledge levels, but this will require that you pre-publish an agenda identifying specifically what will be covered when and what level of expertise is required for each segment. This only works if you have latitude in the timeframe allotted by the buyer and the attendee's schedules are flexible as well. At the same time you are going to have to be more selective about the buyers that you involve in the discussion during your presentation. Focus on the influencers and the decision makers, if the group is large.

The timing of your presentation relative to your competitors.
You may not always be able to find this out, but you should make every effort to do so. Where your product is positioned competitively will dictate where you ideally want to be scheduled. The number of presentations will also effect how you want to be scheduled. Here are some general rules to follow if presentations for all vendors are scheduled over a one week period.

- Small number of competing vendors (3-4) and your product is the competitive leader: go first. The other vendors will have to play catch up because you will set the standard.
- Small number of competing vendors (3-4) and your product is not the competitive leader: go last. You don't want a direct comparison between your product and the competitive leader at this stage of the evaluation. Vendor presentations between yours and the leader will help blur the differences.
- Large number of competing vendors (>4) and regardless of your competitive position: go last. The buyer will have learned a lot about the solutions, but the differentiations will be blurred. They will

remember the last vendor more clearly. It does mean that your presentation must be more clearly and dramatically presented to leave a lasting impression.

If the evaluation is spread out over several weeks, you should try to schedule yourself last if you are the competitive leader. They will remember the last presentation best, plus you will be about to fulfill all the shortcomings of the other vendors.

These general rules may not always apply. It will depend largely on your overall competitive position and the competitive differences between your product and the other vendors' products. Be objective about your competitive position. It's hard to do, but it will pay off in the long run. The vendor in the lead going into the presentation could also effect where you want to be positioned. It is usually best not to be buried in the middle of a series of presentations.

Availability and appropriateness of your company's resources.
There will be times when having a company officer accompany you into a presentation situation will be beneficial. If there will be more than one presentation, save your "big guns" for later in the sales cycle. When you do use an officer, make sure that he/she participates in a meaningful manner in the presentation. If your company officer does not participate and add value, the audience will just wonder why he/she is there. "Doesn't he/she have something better to do?" Let

the officer use his/her own strengths not the standard company line. For instance, if the officer is in charge of R & D, their talk should relate to R & D. If he/she is responsible for sales and marketing, their talk should relate to customers. The officer's contribution should be planned and timed. You don't want this person derailing your presentation by talking about a topic out of sequence or by dominating the presentation.

If you are well into the sales cycle and can arrange for the buyers to visit your company, you should try to expose them to as much of the talent in your organization as you can within a reasonable timeframe. The buyer will already be familiar with your sales cycle players and they should have an opportunity to see that your company has depth and breadth. You should groom non-sales personnel and in fact, anyone you place in front of a buyer. They need to understand the account and what is expected of them once they are in front of your buyer. These value-added players do not actually have to make a presentation. You may want a couple of them to join you for lunch to meet the buyer in a casual setting. However, they still need to be prepared and briefed by you. They should have a message that you want delivered to the buyer or there is no reason to introduce them into the sales cycle.

Summary

Before you can organize your presentation, you have to know what the constraints are. The various timing constraints, location, size of the audience, competitive situation and company resources are all key to pre-planning your power presentation.

Chapter 4

Organizing your Presentation

Careful organization will present you, your company and your product in an orderly, understandable fashion and focus your audience on the objectives that YOU have set. Your first step is to set your objectives for this particular presentation. This, of course, will depend upon where you are in the sales cycle. Be sure that your objectives are clear and that everyone participating from your company understands them. Everything you present must fit your objectives or you will lose focus. Try to limit your objectives to no more than 4. Some examples might be:

- Make the buyer comfortable will me, my company, and my understanding of their business.
- Make the buyer understand the competitive differences in my product and technical superiority of my solution.
- Make the buyer feel confident that my product/service specifically address his/her business requirements.
- Make the buyer confident that our customer support will ensure his/her success.

- Show the buyer how we build quality into our product.

- Show our total commitment to our customers.

- Show the buyer's upper management that our solution satisfies the needs of the users and it will cost justify itself with in a specific period of time.

- Show the buyers how we will make the required modifications and provide proof that we know we can meet their deadline.

- Show the buyer our expertise in their industry, how our product meets those needs and why our customers chose us over the competition.

These examples span a variety of situations, but they should give you an ideal of how to define objectives.

With objectives in hand, you are now ready to organize your presentation. How you organize will depend upon where you are in the sales cycle.

Your first presentation should be more formal in nature than subsequent presentations and should follow this general outline.

- Welcome and Introductions
- Presentation of your Company
- Review of Buyer's Requirements

- Product/Service Presentation
- Summary of your Solution
- Close and Definition of the Next Step

Some guidelines will be reviewed for each component with additional attention spent on the product portion of the presentation later in this chapter.

Welcome and Introductions

In a team selling environment, this is the responsibility of the sales representative. The welcome should be very brief. Basically, you need to thank the group for giving you the opportunity to present your product/service and express your pleasure. You should introduce the other people from you company that will be participating in the presentation along with their credentials. If you have people with you that will not be presenting, introduce them as well. You should clarify who they are and why they are there. For example, they are new employees and are there to observe. Don't let the audience wonder. Also make sure that they will not be drawn into the discussion. Limit the introduction of the person presenting the product. Their detailed qualifications should be presented just prior to you turning over the meeting to him/her. You want to present this person as the product expert and you want the audience to be very aware of this as the person starts to talk.

Your company presentation should be short, relate directly to the buyer's value-chain, be interesting and set the tone for the rest of the presentation. Don't tell them all the things you've already told them. Some pertinent facts about your company, your customers and your products that they haven't heard yet should be sufficient. Be careful about using any of the buyer's direct competitor's that are your customers in front of the group. This can make your audience uncomfortable. That's like serving Coke to the Pepsi bottlers!

Your introduction of the product presenter should establish him/her as an expert. Education, professional organizations, special projects, business experience and experience with the product should all be included. You should show genuine excitement about this person being available to do the presentation. What you are trying to due is place value on this person and value on their time. If done credibly, you make the audience fell confident in this person's ability.

If you are the sales representative and you will be doing the product presentation, you should state your qualifications during the introductions. Watch out for sounding pompous if your qualifications are extensive. Focus on your industry and product experience rather than your sales success. You should also make your company presentation very brief. You need credibility as a product person during this presentation and you don't want to continually remind them that you are a sales representative.

Review the Buyer's Requirements

In a team-selling environment, the product person should conduct this review. This should be a concise review of the needs that you discovered during the survey process. Don't try to list everything. Try to focus on where the greatest needs are and also where your product provides significant solutions. Try to incorporate as much of their terminology as you can. Your goal here is to let the buyer know that you understand their business and their needs. As you cover each item, relate it back to the person(s) that told you about the need. Do not try to offer a solution to the needs at this point.

This requirements list should be organized in the sequence that you will use in presenting your product. This will start orienting the audience toward how you are going to discuss your product. While reviewing the list get affirmation from the audience. Some examples are: "Did I understand that correctly?", "Mary indicated that solving this will save her hours of time", or "Is this a problem now?".

Do not try to solve the problems or explain how your product works during this review. You should focus completely on their problems, empathize with them and show that you understand. If you can segment their needs into logical groupings from an application standpoint, do so. At the end of each logical grouping be sure to ask if any other requirements should be added to your list. By doing this,

51

you are getting agreement in advance that you will be covering their needs during your presentation. This may also mean that you must verbally expand your presentation to include some new topics.

The Product Presentation

The product presentation needs to be the most organized segment of the overall presentation. You will be presenting a new product to the buyer, trying to relate it to the buyer's terminology and matching the buyer's needs to your product solution. If you have any hopes of the audience understanding you and matching it to their individual needs, you will have to present it in a way that they always know what your approach is, where you are and where you are going. A standard rule of thumb is "tell them what you're going to tell them, tell them and then tell them what you've told them."

You should always start with a very high level overview of the entire product, showing its components broken down into logical groupings. By doing this, you are showing the audience the scope of your product and you are showing them the sequence in which you are going to present the product. Each component should be overviewed just enough to let the audience know which category of their needs fit into that particular segment. This ensures the audience that you are going to cover what they came to hear and it will keep individuals from asking questions out of sequence.

The product components should be logical groupings of business activities. They should NOT be how your product is technically structured. For example, it your application is built using relational technology, don't structure your presentation based upon the tables. It's too confusing and it does not relate to solving a business problem in a direct manner. Don't ask your audience to interpret how your product relates to their business. You have to be specific and show them with examples so they can associate your product solution to their day-to-day jobs. The technology that supports your product should be a different component in your presentation and identified as such. If you do not have IT buyer personnel in the presentation audience, save this portion of the presentation until you do or schedule a separate session. Often it works better to have a separate technical session with the IT people so you can be as technical as you need to be without losing the non-technical members of the audience.

Once you have done your overview of your product, let the audience know the sequence that you will be presenting your product and get agreement from the audience that this is acceptable. You should have previously agreed upon this sequence and timeframe with your primary contact prior to the meeting. Do be prepared for some changes here.

You are now ready to move to the next level of detail. Again, you must now overview the component or section of your presentation similar to how you did the complete product. It gets the audience focused and again helps prevent out sequence questions that can be very disruptive to the flow of your presentation. Once you have completed your explanation of the component, summarize (tell them what you told them) the benefits of the component as they relate to the buyer's needs before moving on to the next component.

Product demonstrations present a challenge and unfortunately, many vendors try to sell their product with only a demo. This is a painful way to sell a product and usually not nearly as effective as a combination of presentation and demonstration. It is very hard to follow a presentation on an online screen, panel or page at a time. What ends up happening is the audience does not ever grasp how the product fits together or the overall benefit. Demonstrations should be used as proof statements. They can be very effective if a product presentation precedes the demonstration. Demonstrations should also be brief in duration. Looking at online demonstrations is difficult for the audience even when your screen layouts are crisp in appearance and to you seem easy to read. Product demonstrations interspersed throughout the presentation are very effective if they can be done seamlessly.

Don't try to show every feature in a demo. Try to show some straightforward business flows that the audience can easily relate to their personal situation. Detailed demos, when required, should be done with small groups of buyers to be effective. Then you can focus on the specific needs of that particular group.

Summary of your Solution

The final step of your product presentation is to return to your original requirements list that you prepared from your survey and reviewed at the beginning of the presentation. This time you want to go through the list matching your solutions to their needs. This allows you to sell the product by using their own words and showing them how your product meets their specific needs.

This approach may appear repetitious. It is! And that is what you want. The more times they hear it, the more likely they are to remember it.

Close and Definition of the Next Step

In a team-selling environment, the sales representative should do this. Your close should be brief. It should contain a "thank you" and a statement to say "We want your business". It should wrap up any outstanding issues. Some examples might be: "We will send you sample reports tomorrow"; "I will find the answers to the questions that are outstanding and call you on Friday"; "I will have our technical expert call you Monday to schedule a conference call". Be specific and follow-up when you say you are going to. Your finale is to get a commitment on the next step. This might be another meeting, a client visit or any event that will take you closer to getting the contract. Don't leave the presentation without getting a commitment to the next step!

Summary

The organization of your presentation can mean the difference between winning and losing the business. Your audience must be able to follow your presentation from beginning to end and understand the message that you intend to deliver. If it does not follow a logical flow from the buyer's perspective, you are leaving too much to chance. By keeping the buyer's perspective in mind, you will present a solution that they can relate to and appreciate.

Chapter 5

Visual Aids

This chapter discusses various types of visual aids and includes some tips on how and when to use them. A visual aid serves only one purpose during your presentation. It helps your audience visualize what you are saying. This is especially true in technology product sales presentations. For example, explaining a relational database or a complex SAN without visuals would take a considerable amount of time with no guarantees that the audience would understand the concept. By using a visual aid, in a manner of seconds the concept is easily understood. The same is probably true for your product.

Visual aids should not be used to step you or your audience through your presentation. Unfortunately, this is their most common use and as a result they are usually less than effective. Keep in mind that you do not need a visual aid for everything you say! A visual aid should be used to enhance your audience's level of understanding.

Your visual aids will be heavily dependent upon the size of your audience and the setup of the room. Just remember that everyone in

57

the room should be able to easily see and read your visual from anywhere in the room. If your audience is straining to see your visual, they are not listening to what you are saying.

Visuals should be simple. If they are cluttered or too wordy, they will not be effective. You are probably trying to cover too much material with one visual.

If you plan to use visuals, and in most cases you will want to use them, be sure that you are using a medium that is comfortable for you and that does not require much effort on your part. You want to be focusing on your audience, not trying to manipulate your visuals. Variety in your media is usually best. I personally like giving a presentation in a lighted room so that I can see my audience and their reaction to my presentation. I find that people's attention span is greater in a lighted room and it is much easier to get the audience involved in what I am doing. Therefore, I stick to media that allows me to do this. Overheads, flipcharts and whiteboards all lend themselves to lighted rooms. PowerPoint or equivalent presentations need to be specifically designed to work in lighted rooms. Online demonstrations do not lend themselves to lighted rooms, but these will be discussed later.

Company prepared videos and interactive videos are terrific visuals to intersperse throughout your presentation if they are available and appropriate to your material. First presentations are an excellent time to use them to introduce your company and its philosophy. Videos will not be covered, as presenters do not usually have control over the content. We will focus on visuals under the control of the presenter.

Let's look at various types of visual aids and how you can most effectively use them. Some specific guidelines for visuals will be provided at the end of this chapter and they apply to all media types.

Slides

Slides, if they are done well, give your presentation a very professional, polished look. However, they are the least flexible media and they typically require you to do a presentation in a darkened room. If your chosen medium is slides, you will need a means of customizing the presentation at the last moment. Back in the days when I used them, I color coded the top of the slides to allow me to remove or rearrange slides just prior to the presentation when required. It also allowed me to skip sections of the presentation on the fly when the need arose.

You do want to avoid giving a presentation in the dark! If you will be presenting in a room where you can control rows of lights and you are using slides, you can darken just the front of the room to allow the slides to be seen without plunging your audience into the dark. This will also allow you to have contact with your audience

Sides are best used in seminar situations when you are doing a general presentation and you are not selling to a specific buyer. Slides, like PowerPoint presentations, do offer the ability to have high quality "fun" slides to liven up seminar presentations.

Tips for using slides:

- You will need to memorize the sequence of your slides. This will enable you to position yourself well in front of your slides and out of the darkened portion of the room. Your audience needs to be able to see you in order to relate to you. You do not want to be turning around to see where you are in your presentation and which slide you are on. In an ideal situation, you screen should be placed in a corner, which would allow you to glance sideways at the screen. Unfortunately, most of the rooms you will be presenting in will not allow this flexibility. Be prepared.
- Tuck the clicker cord into the back of your waistband. This will keep the excess cord in back of you and help prevent you from stepping on the cord and jerking the clicker out of your hand or even worse tripping. To

ensure the proper length of cord between you waistband and your hand, hold the clicker in the hand that you will be using and hold your arm out parallel with the floor. Tuck the cord into your waistband while in this position. You should have enough cord to still comfortably move your hand with the clicker to make gestures during your presentation.

- Color-code the top of your slides into logical modules, functions or specialized features. This will allow you to quickly customize your presentation prior to speaking or easily skip sections during the presentation should you have a need.

Electronic or PowerPoint™ Presentations

Electronic presentations are the most widely used and abused today. The logistics of a PowerPoint or electronic presentation are the same as for a slide presentation. You can't of course color code within a PowerPoint as described for slides. You can however develop a highly organized presentation with clear transitions between topics that are easy to find should you need to skip a section. You do not want to "click" through all of the undesirable pages. You should change modes and click directly to the section you want.

Because of the graphics capabilities within electronic presentation tools, it is relatively easy to overwhelm your audience with too much information, too many graphics or just too much stuff on one page or screen. You need to resist this temptation. Because of the complexities of today's solutions, complex graphics can quickly and accurately show the complete solution. There are some pitfalls. Presenters have a tendency to present too much too quickly and overwhelm the audience. As the presenter, you are extremely knowledgeable about your solution, but your audience is not. It is important to continually remember this. Complex diagrams and graphics should be eased into. You should start with very high level diagrams and gradually become more granular with your solution. Take caution not to cover too much complex information at one time. You should present your solution in smaller less complex components, gradually building the entire solution. It is a good idea to test your presentation on someone that is not an expert in what you are presenting and listen to what they have to say. Keep things in perspective. You are not there to show everyone how smart you are or how much more you know than your audience does. You are there to sell your solution in a clear, concise manner and in a way that your audience will understand. Even if your company is going to install the solution, the customer has to maintain it. Don't scare them off!

In a managed services situation or an ASP, you should not be presenting the technology that the customer will never see. Present only the part of the solution that directly relates to what the customer is expected to do and the results that you will provide. They don't need to know the complexities of how you will provide the solution.

Overheads/Transparencies

Overheads are probably the most flexible media while still being very professional for product presentations. They should be created as a backup for PowerPoint presentations to cover technical problems with your laptop or display device. Like PowerPoint presentations, they allow you to customize each presentation. If you use clear transparencies and dark colored print, you should not have problems using them in a lighted room. You should not use colored transparencies. They just don't look that good. Colors for your print and graphics should be tested in a lighted room to be sure that everything on the overhead can be easily read from the back of the room. Again, you don't want the audience straining to see your visual.

Tips for using overheads:

- Don't cover up part of the overhead as you start to use it. This procedure can be very annoying to your audience. It's a form of teasing that is not generally well accepted. Let the audience see the whole page. If you are trying to build on a concept, you should be overlaying the overhead with additional overheads.

63

They can be taped together so that you just have to lift each overhead and you can be assured that each one will be lined up properly.

- If you do not use cardboard frames to you overheads, you will find that the overheads to not stay straight on the projector. Static electricity makes them slide around. Place a coin on one corner and this will hold the overhead where you want it to be.

- If you are not using the visual, turn of the projector. Do not leave an unrelated overhead up in front of your audience. This is distracting. You want them to listen to what you are saying, not thinking about what is on the overhead. Remember, you do not need a visual for everything that you say.

Whiteboards and Flipcharts

These can be extremely effective during your presentation to illustrate a specific point not provided on a prepared visual. If you do not have a neat handwriting, you should not depend on this media and should limit your use of it. If your flipchart looks messy, it inadvertently reflects on your professionalism. You should also not write on a flipchart or a whiteboard unless you are trying to illustrate a point. I have watched presentations where words were written down and 20 minutes later, I couldn't figure out what point the speaker was trying to make. When you are finished with the flipchart page or making a point on the board, turn the page or erase the board. Don't leave distracting notes in front of you audience.

Tips for using whiteboards and flipcharts:

- Use abbreviations or acronyms ONLY if you are positive that your audience knows what they mean.

- Place the board, if it's movable, or the flipchart in a position that you do not have to turn your back on the audience as you are writing. Do not talk while you are writing. Write. Then, talk. Otherwise you will be talking to the board or flipchart, not to your audience. For fixed boards, you may have to turn your back to the audience. Keep your writing brief. Write. Then, talk.

- If you are using a prepared flipchart, use only every other page. This will ensure that information from the next page does not show through on the current page. Leave blank pages for additional comments or notes in regular intervals throughout to give yourself flexibility. Staple each printed page to the following blank page. This will give you something more substantial to grasp as you turn pages.

- Always print. Your handwriting will always be harder to read for someone who is unfamiliar with it. It is also easier to print in a straight line than it is to write.

Online Demonstrations

Online demonstrations, as mentioned earlier, should provide proof of your product and should not be used exclusively to sell your product. Demonstrations should be presented as business process flows rather than feature, feature, feature. The audience needs to be able to relate to what you are showing them.

Be sure that your demo can be seen from the back of the room and that the focus is as sharp as possible. Some equipment may create shadows that can be very hard on the eyes. Even when the projection image is clear, demos are hard on the eyes. Online screens, panels and pages are designed to be used by a single user. They are not designed to be viewed by a large audience and from a distance. Whenever possible, demos work best with someone on the keyboard other than the speaker. This requires practice and it should be understood that the speaker is in control of where the demo is going. I have seen many, many of these demos where this was not the case and it's not pretty! By using two people, you allow the speaker to focus on the audience and not on the demo.

Prior to the demo, you should get specific examples from the buyer. If at all possible, use some of their data. If they recognize their data in your product, they will more readily identify with your solution and start thinking about it as their solution.

A request for a completely customized demonstration from a buyer is quite common today. I personally would try to resist doing this. It is time consuming and expensive for the vendor and there are no guarantees that a fabulous demo is going to win you the business. The other disadvantage is the buyer's script may not adequately demonstrate your product or may not show it in its best light. This really needs to be a company policy and something your company is willing to finance. In some cases, companies are providing customized demos for a fee.

General Guidelines for Developing Visual Aids

- On list visuals, limit the number of lines to 6 or less, plus the title. You don't want your audience getting ahead of your presentation and you want your visual to have a neat crisp appearance, not cluttered.

- Limit the number of words per line. If a topic takes 2 lines, it's probably too long – cut it.

- Use plain text rather than script or hard to read fonts.

- Use graphics, drawings, pictures or diagrams whenever possible

- Cover only one concept per visual. Keep your audience focused.

- For colored visuals, develop a conservative but cheerful color scheme and stick to it for consistency.

General Guidelines for Graphics and Diagrams

- Less is more.

- Gradually build up to complex solutions.

- Do not overwhelm your audience with too much detail at one time.

- Test your graphics and diagrams on a non-expert audience before using.

Summary

Visual aids can significantly enhance your presentation if used properly. Visual aids can also detract from you presentation if they are poorly designed, ineffectively used or overpower your product. Your visual aids reflect on YOU the presenter, so you want them be professional and representative of the quality of your products/services. They are not a testament to your technical prowess.

Chapter 6

Physical & Professional Environment

The setup of the presentation room can have a dramatic effect on how well you can control your audience. Naturally, you may be limited in your flexibility by the number of attendees and the facilities you have to work with. This chapter will discuss various room setups and how to take advantage of them. The purpose of this chapter is to help you establish the right physical and professional environment for your presentation.

Physical Environment

We will start with the physical environment in the presentation room. Several room setups are available to you. Their use will depend on the size of the room, the size of your audience and the available tables and chairs. They are listed from most desirable to least desirable for the speaker and the audience. The first four provide a writing surface for the audience and in general a more comfortable environment.

- U-shape
- Herringbone
- Schoolroom

- Conference (qualified)
- Modified Theater
- Theater

U-Shape

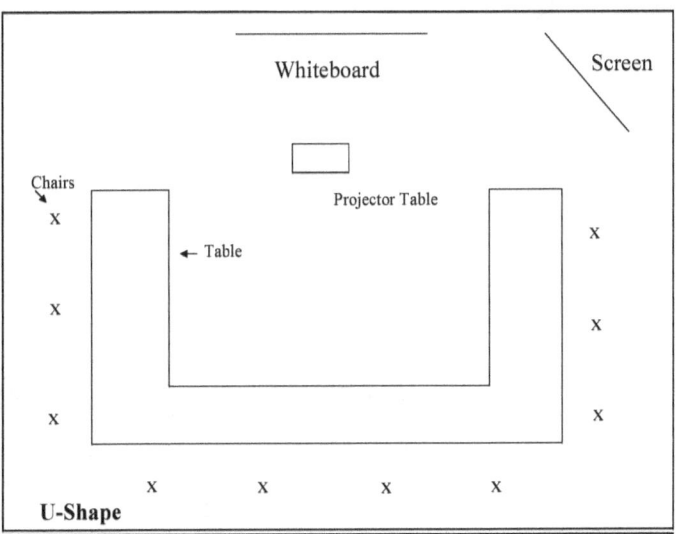

Ideal Size: 6-18 buyers

The u-shape is ideal for the speaker. It gives you easy access to everyone in the audience and provides for maximum control by the speaker. It also allows for open discussion in a comfortable fashion for the audience. In addition, you will find that fewer people will doze off or be distracted since everyone can easily see everyone else. With a small number of buyers, the buyers should be seated on the legs while additional participants from your company should be seated along the back. Do not use up a place on a leg for the speaker.

Herringbone

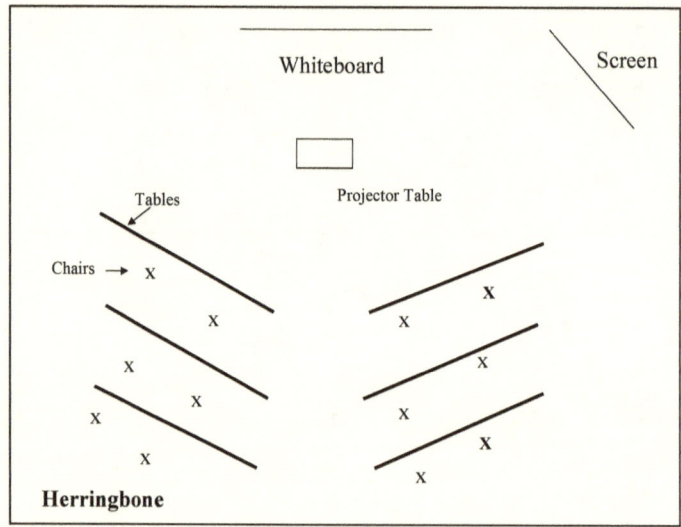

Ideal size group: 2 – 4 people per table, 3 – 6 rows

Herringbone is a modification of the schoolroom setup. It allows for more interaction within the group, because of the angles of the tables and the fact that it is easier to make eye contact across the group. It also provides the speaker with easier access to the audience therefore more control over the audience. Herringbone usually requires a wider room than schoolroom. Again, additional participants from your company should sit in the back or on the outer seats.

Schoolroom

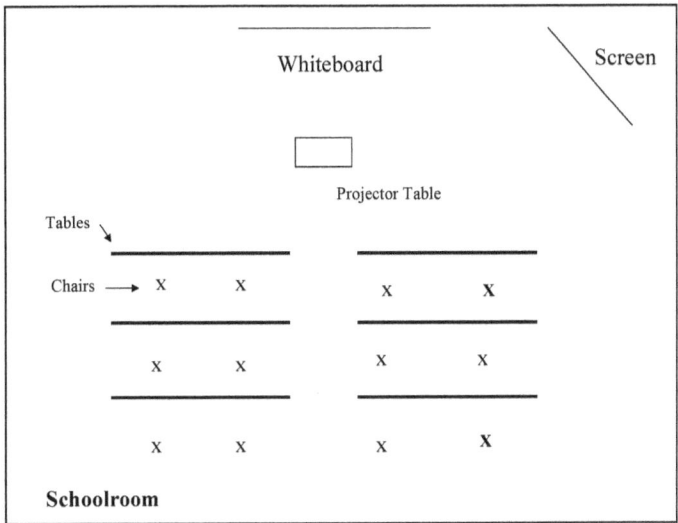

Ideal size group: 2 – 4 people per table, 3 – 6 rows

There are advantages and disadvantages to the schoolroom setup. This setup provides a writing surface for the audience and it is a familiar learning environment arrangement. The disadvantages, however, outweigh the advantages. This setup does not allow for a comfortable flow of interaction between the audience members. Everyone except those on the back row have to turn around to engage in any discussion. It also means that people are lines up behind each other interfering with a clear line of vision to you and your visuals. The diagram above shows the projector screen across the corner. This is ideal for you and your audience, but in most cases, the screen is installed and is in the center of the wall.

Conference

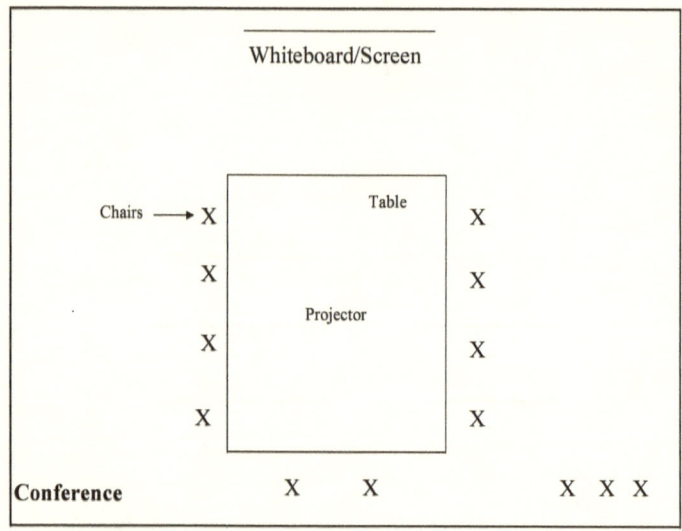

Ideal size group: 6-12 people

This is the most common room setup that you will encounter. Unfortunately, it is also the most difficult setup for you to have complete control. If multiple people are in attendance from your company, make sure that they sit at the far end of the table from you or along the sides of the room, forcing the buyers to sit on the legs of the conference table. If the table only has enough chairs for your buyers, your company people should not sit at the table, but along the sides. This will allow them to help you regain control should you lose it. About the only changes that you can make in this situation are 1) remove any extra chairs from the room so that you can control where people sit and 2) move the table back from the front of the room giving you as much "moving around space" as possible. That is assuming of course that you can physically move the table.

I place conference style room setups as a qualified fourth only because it does provide a writing surface for the audience.

Modified Theater

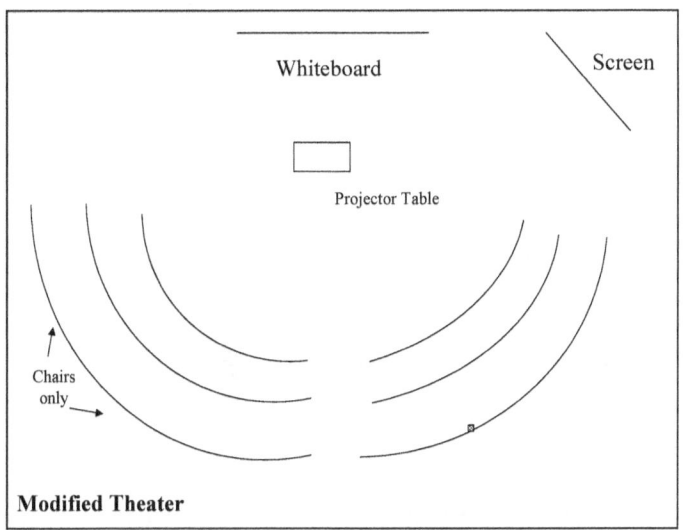

Ideal size group: greater than 20

This modified theater is theater style, but with curved rows. Although this setup will make open discussions easier than with theater style, you should follow the same guidelines as theater style. You don't want to lose control. This set up is easier on your audience and therefore preferable to use over theater. This setup should only be used if you have no other alternative.

Theater

Ideal size: greater then 20

Theater style should only be used if you have a very large group and no tables available. This is a very uncomfortable setup for you audience and makes it very difficult for them to take notes during your presentation. If you have to use this type of setup, be sure to take frequent breaks even if they are just "stretch" breaks. It will also be hard for you as the speaker to control the audience, because you only have access to the front row and the first few isle seats. It will be difficult to gracefully regain control if you should lose it.

Your choice of room setup will be heavily dependent upon the situation that you are presented within the buyer's setting and the size of your audience. You may be limited by the buyer's environment as to what you can change or adjust, but where you can, do it. In terms of the best room setups for YOU the presenter, they are in priority sequence: u-shape, herringbone, schoolroom, conference, modified theater and theater.

You are probably thinking, "The room is already setup when I get there". That's true in most cases. But, it is important for you to remember that it is YOUR presentation and you have the right to change the room setup if it does not meet your needs. If you are to do your best, you must take advantage of every opportunity to put your best foot forward. The room setup will have a dramatic impact on your ability to control your audience. It will also enhance your presentation if your audience is comfortable and detract from it if they are not.

I have mentioned controlling your audience several times in this chapter. We will be discussing this in detail in Chapter 8, Controlling and Analyzing your Audience. These activities happen during your presentation. Establishing the physical and professional environment happens prior to the start of your presentation.

Professional Environment

Setting the professional environment really starts during your survey. You are establishing your credibility to some degree, but mostly you are starting the process of understanding the buyer's needs and showing empathy for their problems. Being the most prepared vendor will also impact the initial professional environment when your buyers enter the presentation room.

You should have arrived early, changed the room setup if necessary and possible, checked out all of your equipment and placed any marketing materials in their appropriate place. Prepared name tents are a good idea and I recommend them. This helps you place people where you want them to be and prompts you on their names during the presentation. Nametags are only appropriate if the presentation is held in your company's offices or in a seminar situation. When you host a buyer presentation in your company's offices, you want to be sure that the employees not involved in the presentation be alerted that a buyer is in fact a buyer. In a seminar situation, nametags help you and the attendees interact.

Marketing materials should be used sparingly pre-presentation. Don't put any materials on the tables unless YOU are going to use them. You may choose to provide a copy of your presentation, but remember that some people will try to get ahead of your presentation. If you prepare an information packet of information, be sure to review it prior to your presentation and limit the information provided. You don't want your audience reading the packet of information rather than listening to your presentation. You should always have an agenda. I recommend having a presentation folder with an agenda and possibly some information on your company. The audience will then have a place to store any additional information that you may pass out during your presentation. People will usually wait to read company information if they are interested in your product and it should not interfere with your presentation. Brochures, sample reports, graphics, etc. should be handed out during the presentation to illustrate a point that you are making at the time. Make sure that you have these organized and enough copies for easy dispersal.

In very large group settings (i.e. over 20 people), prepare packets to hand out at the end of the presentation. This does not apply to seminars. See Chapter 9, Seminars.

In addition to name tents and marketing materials, you should also be prepared mechanically. If you are using slides, always carry extra projector bulbs even it you won't be using your own projector. If the bulb goes out, you can quickly replace it and go on, eliminating a time consuming scramble for a new bulb. Always bring your own markers for the whiteboard or flip chart and make sure they are still moist. Bring your own pointer. If you are using a computer dependent presentation (i.e. PowerPoint), bring a back up such as overheads in case of technical issues. The same applies for slides. Always be prepared to use a whiteboard or flipchart in the case of emergencies. I did a presentation once in a storeroom on a flipchart when I had come prepared to do a slide show. Be prepared for the unexpected.

If you are well prepared, well in advance of the buyers entering the presentation room, you will set the stage for a professional presentation. You will feel confident and more relaxed, because you have control over the situation. Your preparation will be noticed by your audience and appreciated. It will be a relaxed, professional setting for your buyer!

Summary

You want to place your buyers in a comfortable room setting with an atmosphere of professional calmness to make them receptive to your presentation. If your audience is physically uncomfortable or you appear unprepared, your audience will not be emotionally ready to hear your message. You do this by selecting an appropriate room setup whenever possible and by being well prepared, well in advance!

Chapter 7

Presentation Delivery

This chapter discusses the actual delivery of your presentation. During Chapter 4, Organizing your Presentation, several techniques were describes that will be part of your delivery. This chapter focuses purely on the delivery and assumes that you have followed the recommendation in this book in preparing your presentation. It is primarily in the form of tips for being a successful power presenter.

<u>Make it interesting</u>

You will be the most successful if you deliver your presentation as if you were telling a story. It should be conversational and easy flowing. You should use interesting stories to explain a benefit to a feature. Customer successes are always best. It gives you an opportunity to inoffensively brag about your product through your customers' experiences. Customers that have been successful with your product/services will naturally make you feel proud of your product and this will be reflected in your voice. Success stories of customers will instill confidence in the buyer for your product.

Humor is also an excellent tool for making a presentation interesting. Keep in mind that you are selling a product, but your audience also expects to be entertained to some degree. You want the audience to have a good time as well as want to buy your product. Humor is tricky and should be used cautiously. NEVER tell off-color jokes or stories, political jokes or religious/ethnic stories. Make sure your humorous story or joke relates to what you are saying. Always use a type of humor that 1) you are comfortable with and good at and 2) fits your personality. Not everyone is good with humor; so don't feel that you have to use it. Personally, I am a terrible joke teller so I never tell one. I am much better at one-liners in reaction to a situation. Test your humor on non-buyer audiences before taking it into a buying situation.

Another way to make your presentation interesting is to add life, fun and human interest. An example that I used is describing an enthusiastic customer project team installing our company's product. This group was so "fired up" that they went out and had t-shirts made identifying them as our company's product project team. On milestone days they would wear them to work. Another example I used was when no one was sitting up front when I had a large group. I would say, "I can't believe that you are all sitting back there in the $5 seats when we've got all these $10 seats vacant. Come on down, live a little!". I never had a situation when people didn't move to the front. Poking fun at yourself may work as well. At a large conference one time I was presenting and could not for the life of me pronounce "ergonomically". I laughed and finally got it out. The audience laughed because I did. Later in the presentation, I had the exact same problem. It was one of those days and after a couple of tries, I laughed and said, "Oh well, you know the word". Again, the audience laughed.

You will need to think back over your own experiences to come up with customer examples, personal situations, recent articles or any memory that may add some spice to your talk. Again, be sure that your "spice" relates to your topic. You will need to add some spice at least every 10 minutes or so.

Interact with our audience

Using your survey information, you should have pre-planned questions. DON'T EVER ask a pre-planned question unless you know the answer or you may plant one of those land mines discussed earlier. You can also ask a question to buyer, giving yourself time to formulate an answer by asking, "How do you do that now?", for example.

You will need to get feedback from the audience throughout your presentation such as "Does this solve your problem?" or "Isn't this one of your concerns?" or "I know for XYZ customer, this saved them 8 hours of time. Could this save you time?"

Another way to get the audience involved is to get one of your buyers to provide you with an example. It might be one that you heard during the survey process. You might say, "John, you had a really good example of this, but I don't think I can remember it just right. Could you tell me again?" This gets your audience involved. You want your audience involved. They are going to remember more of what they say than what you say. It's unfortunate, but true.

Remember: *You are trying to sell them a solution to THEIR problems. The problems are THEIRS. Let them talk about them and join in the discussion on how to solve them. Make them feel that its THEIR presentation, not yours. That is the way that you want your buyers to feel. I believe in planting the idea, prompting the buyer for the solution and then, congratulating them for making a good decision.*

If you are having a hard time getting your audience involved, keep asking open-ended questions. Open ended questions start with "how", "what", "when" or "why". If you are having too much discussion, ask closed ended questions. Closed ended questions can be answered with a "yes" or a "no".

If you don't understand a question from the audience, don't make a guess at what you are being asked. Nothing is worse than getting the answer to a question that wasn't asked. This is extremely aggravating to the questioner and could prove embarrassing to you. Always take the burden on yourself for being the one with the communication problem. You should acknowledge that you don't understand the question and ask him/her to restate it. If you still don't understand, ask him/her to give you an example. If you still don't understand, postpone the question and schedule to answer it during a break. Often times someone else in the audience will help you out by restating the question in a way that you will understand. If no one does, it is probably not that important to the audience as a whole.

Being in sync with your audience

You will need to constantly watch our audience to make sure that they are following your presentation. To ensure that your talk is meeting their expectations, you should be watching for their reactions to it. You've got a problem if:

- People are talking among themselves.
- No one is taking notes.
- You are NOT getting questions from the audience.
- People are yawning.
- You are NOT getting nods of agreement.
- People are avoiding eye contact with you.
- There is a general lack of interest and participation.

If this happens and it can, you should stop the presentation. You should take on the burden for creating the problem, even if you are doing exactly what you were told they wanted. The buyer's internal agenda could have changed without your knowledge. When you stop the presentation, you should acknowledge to the group that you seem to be off track. For example, "I don't seem to be talking about what is of interest to you. Did I misunderstand what we were to cover today? Let's take a moment and discuss what you would like to hear?" There may be times when the area to cover, due to a change in the internal agenda or general miscommunications, that you are not prepared to discuss. If this should happen, you should not try to "wing it". You may end up being more destructive to the sale cycle than helpful. Don't waste any more of their time or yours. Reschedule the meeting with the appropriate resources. This should rarely happen, but it could.

Product Knowledge

It is mandatory that you thoroughly know your product. You do not need to tell everything you know in each presentation, but the more you know the better off you are. Many times a feature of your product will perform a function that it originally was not designed to do. If you know not only the features, but also how the product works, you will be able to apply this knowledge to the benefit of your buyer.

If you are weak on certain parts of the product and you know they will come up during the presentation, be sure to take appropriate resources with you. If you sound knowledgeable during most of the presentation and flounder through a particular area, it can undermine your credibility on everything else that you have said. If this area of the system warrants a separate presentation, you should explain this up front. Then your audience will not be expecting you to be knowledgeable about this topic.

If you do not know an answer to a question, do not make one up! It is much better to say, I don't know, but I will be happy to find out for you". Then make sure that you do. Write it down in front of your audience. This makes sure that you don't forget, but more importantly, it makes the buyer feel that their question was of value. If the evaluation is dragging out and you need to create reasons during the presentation to get back into the account, pre-plan some "I don't knows" ahead of time.

Never make commitments that you are not confident that your company can deliver. You must always remember that they are a buyer today, but they will be a customer tomorrow. You will need them as a reference. Negative references can have a serious impact on your ability to sell your product.

If you set an expectation that your company cannot meet, you have set yourself up for a bad reference. Your good references are not going to typically go around spreading the good word on your product. Bad references, however, cannot wait to tell someone.

Listening

Being an effective listener is one of the hardest skills to master. This is especially true the longer you do this job. You will hear the same questions over and over again. The natural tendency is to answer the question before your buyer has finished the question or start thinking about your answer and tuning the buyer's question out. You should never interrupt a buyer during his/her question. It's rude. If you tune out the buyer question, you may end up answering the wrong question. Keep in mind; it's not an "old" question to the buyer. Look directly at the buyer asking the question, approach him/her if you can, nod in agreement if you understand the question and force yourself to listen. Focus exclusively on this buyer until he/she has finished speaking.

Terminology

During your survey, you should have collected a number of the buyer's terms. Be sure to use these during your presentation. If you don't know their terminology, try to use generic terms. Your internal terminology, even if you explain it, will force your audience to interpret what you are saying during your presentation. If you do this, you are expecting too much of your audience. They won't do it. And, you won't clearly deliver your message.

Summary

An interesting presentation from the buyer's perspective with active involvement by the buyers will ensure that your audience will remember you and your presentation. If the buyer can relate to what you are saying, you have a much higher probability of winning the business.

Chapter 8

Controlling & Analyzing your Audience

Controlling and analyzing your audience are the final steps to becoming a power presenter. You have to be in control throughout your presentation. The previous chapter described how to get prepared for you presentation and how to deliver it. At this point, you have most of the information available to you to prepare to control and to analyze your audience. This chapter will deal with what you know at this point and also how to handle unexpected situations during your presentation. Of the two, control and analyze, control is the most important. Control will be discussed first.

Controlling your audience

Controlling your audience will maintain the flow that you have pre-established and make the experience more beneficial for the buyer. You have probably been in presentations or meetings where the leader was not in control. In this situation, the purpose can often get lost or the subject matter never gets covered to the satisfaction of the attendees.

The same will be true in your presentation. Each person in your audience has an expectation of your presentation and the most important one to each buyer is that you tell him/her what he/she needs to know about your product. For each individual, the need may be different dependent on the role he/she plays in the buyer's company. If you lose control and you do not meet each person's needs within reason, you have not done your job as well as you could have. Several techniques are available to you to help you control your audience and your presentation. A well-organized presentation that is presented in an orderly fashion and explained to the audience along the way, as discussed earlier, is a big step in helping you.

Now we will discuss some techniques that will help you during the presentation. They include:

- Your body language.
- People placement.
- Flow of conversation.
- Taking charge.

Your body language

Keep in mind that we are discussing control in this chapter. Obviously, you will be using a lot of body language throughout your presentation. What we want to concentrate on here is controlling someone in your audience. You will experience situations where someone is 1) talking too much, 2) giving you a hard time, or 3) having a problem with someone else in the audience (buyer internal personality conflicts that are disruptive to your presentation). Your number one rule should be to avoid asking one of these people an open-ended question. By doing so, you are giving them what they want, a platform to take over the presentation. However, this is not always possible, or they may just talk anyway. There are two other ways that I have used successfully that are less obvious and usually work quite well.

The first is entering the buyer's personal space. Everyone has personal space whether they realize it or not. Think about the last time you met someone for the first time and they got too close to you or you just felt uncomfortable for no apparent reason. They probably entered you personal space. For most Americans, a 3-4 foot radius is considered personal space. Other cultures may have different personal space criteria. You need to be aware of this if you present internationally. There are other cultural differences as well that you should be educated on based upon the country where you will be presenting.

When someone enters your personal space, you will feel uncomfortable if they are someone that you are not close to. YOU as the presenter are not going to be close to anyone in your audience, so you can take advantage of these feelings. When someone is being annoying, move into their personal space. Because they feel uncomfortable, it will usually make them be quiet. You will have to ease into their space, not march into it, and you should position yourself so that you do not have direct eye contact with this person.

The second technique is to cut the buyer off from direct eye contact with you. You can do this by positioning yourself where the buyer cannot easily get your attention. This too needs to be done subtly. You can position yourself by his/her side or just at an angle where eye contact is impossible. If you ease into this position, the buyer will probably not realize what you are doing.

Both of these techniques can stop a buyer from taking over your presentation. Neither is offensive. Both help you remain in control or regain control if you temporarily lose it.

<u>People Placement</u>
Your survey should have revealed some information on the personalities that will be involved in your presentation. You will probably have a mix. There are only 3 types of individual that you need to be concerned with. They are:

- Personality problems within the buyer group.
- Potential trouble makers for you (overtime talkers or negative buyers)
- Potential supporters.

Controlling where certain buyers sit can dramatically improve your chances of controlling your audience. There are two ways to get people to sit where you want them. One is by name tent. However, if the presentation is on the buyer's site, they will probably not hesitate to move the tents. At your company's site, they will be more reluctant to do so. The second and more secure way is to walk them to where you want them to sit. If you have personality problem buyers or troublemakers, don't put out your marketing materials. When the buyers come in, take your materials and while talking with your problem buyer, walk them to the seat where you want them to sit and place the materials there. In most cases they will sit where you leave them. Put their name tent down at the same time. Don't leave them alone, suggest they get coffee and walk them away from the table leaving your marketing materials and name tent where you want them to sit. If too many people come in at one time, make sure that you have prioritized your problem people and take care of them first. In a team-selling environment, make assignments so that each "problem" is covered. Depending on the composition of your group, this may be a time when you get a "friend" buyer to help by suggesting they sit next to one of your problem buyers, but don't explain the real reason that you need them there.

The placement of problem buyers will depend upon your room setup. Let's take a look at specific room setups and where you should ideally place problem buyers. Personality conflicts within your buyer group can be very disruptive. It may not happen often, but you should always be prepared.

We will look at the room setups discussed in Chapter 6, Physical and Professional Environment.

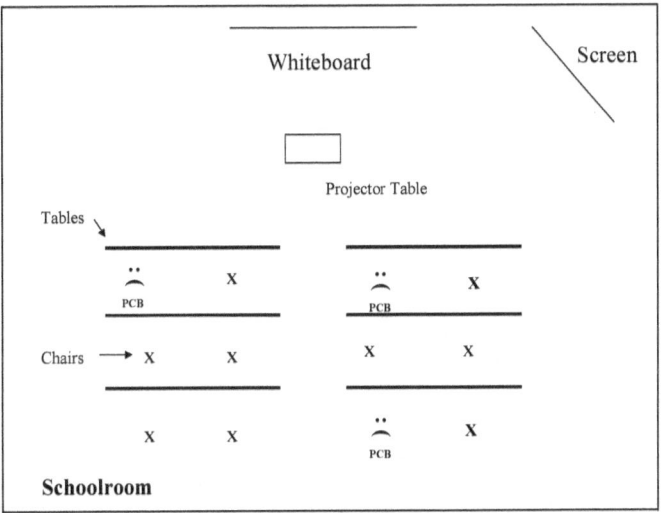

In this schoolroom setup, the personality conflict buyers (PCB) are identified with a face and "PCB". Your goal in this situation is to make sure that personality conflict buyers are positioned where they do not have easy access to one another. This example shows two PCBs on the front row and one on the back row. The front row PCBs are under your control, but they would also have difficulty arguing with each other. First of all, they are up front in full view of the rest of the group and secondly, they would have to argue over two people.

If being up front doesn't stop them you can always ease yourself into the aisle between the rows, cutting off their view of each other. The front row buyers would have to turn completely around to argue with the back row buyer. The back row buyer will be looking at the front row buyer's backs making it difficult to get their attention.

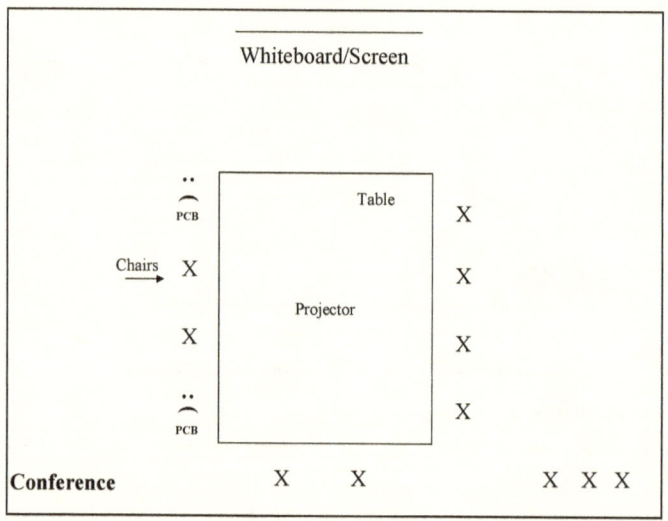

The personality conflict buyers (PCB) are indicated by a face and "PCB" in this conference room setup. A conference room setup is the most difficult one in which to control PCBs. Your #1 rule should be, never let them sit across the table from one another. If you do, you will not be able to control them. In this example, the PCBs are positioned so that any confrontation will have to be done over 3 other people. This alone should be a deterrent.

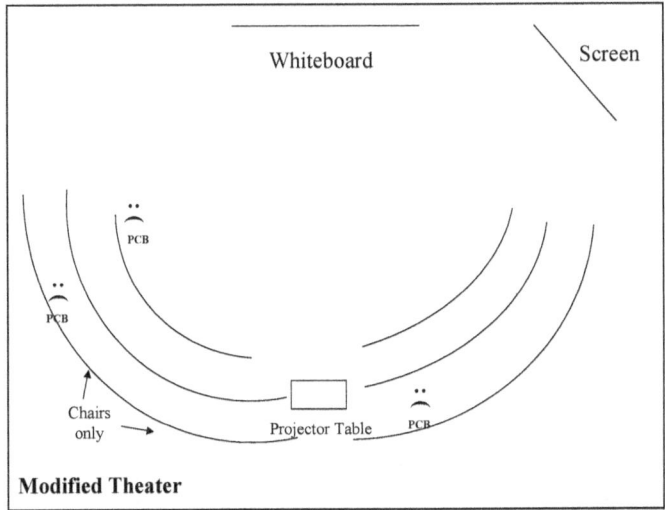

In this modified theater style setup, the PCBs are indicated by a face and "PCB". Again, we have positioned the PCBs so that it is difficult for anyone of them to have access to another PCB. On the left side, the front row buyer would have to turn completely around to access the back row buyer on the left. The back row buyer on the left is looking at the front row buyer's back. The PCB on the right is blocked from the other PCBs by the projector. In this setup, place your most troublesome buyer on the front row where you can control him/her.

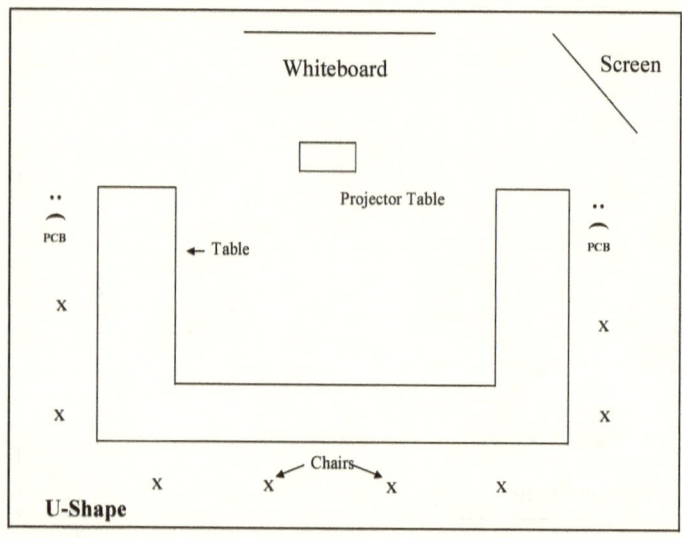

With this u-shape setup, the PCBs are indicated as before. Although the u-shape setup is

preferable to most others because it allows open communication for you and the group, it does cause somewhat of a challenge when you have PCBs in the audience. Place your PCBs at the end of each leg. Hopefully, you won't have more than two! This does allow the PCBs to have easy access to each other, but they will have easy access no matter where they sit in this arrangement. However, you will have them both up front and therefore can control the situation. In this setup, it will be very easy for you to cut them off by standing just inside one of the legs. This will cut off the PCBs easy access to each other and because you are in one of the PCBs personal space, at least one PCB will be neutralized.

In the majority of your sales situations, personality conflict buyers will not be your primary problem. Troublesome buyers will be. There is almost always someone who talks too much or has a negative attitude. In these situations, you always want the troublesome buyer (TSB) up front. Let's look at some example room setups.

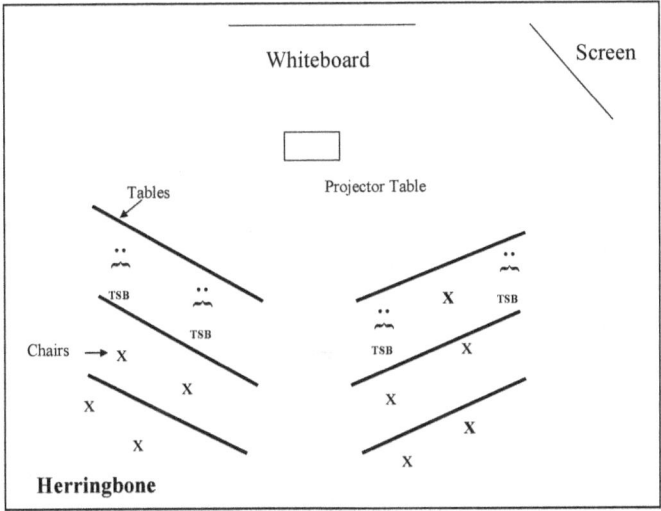

In the herringbone diagram, the troublesome buyers are indicated with a face and "TSB".

You will always want your troublesome buyers (TSBs) up front where you can control their actions. This means more work for you, but if you don't have easy access to them, you will have no way to control the situation. Depending on the number of people per row, try to have one non-troublesome buyer in between two TSBs. Oftentimes a neutral or positive buyer will help you control a negative one. In this particular room setup, it is relatively easy to position yourself to cut off a TSBs eye contact with you without turning your back on him/her.

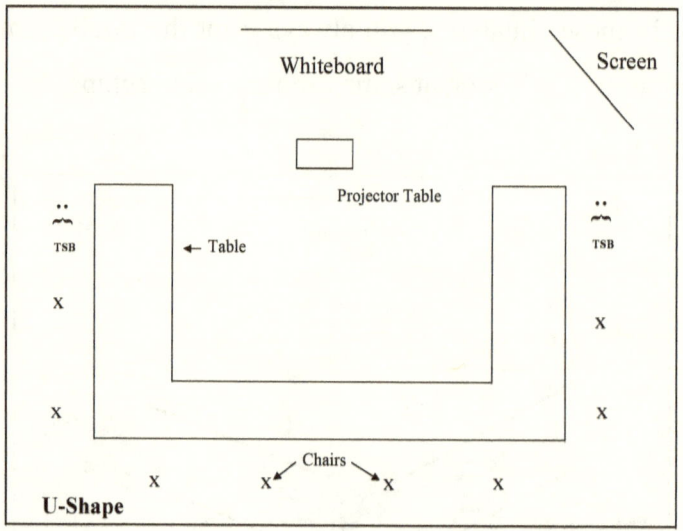

The troublesome buyers (TSBs) are indicated by a face and "TSB". Again, we have the troublemakers up front where you can control them. In this u-shape setup, you will only have control over one at a time, but you don't want them sitting together. They will only feed off each other's negative attitude. In the u-shape it is easy to step into one to the TSBs personal space by standing just inside one of the legs. It also allows you to avoid eye contact with both TSBs in this example.

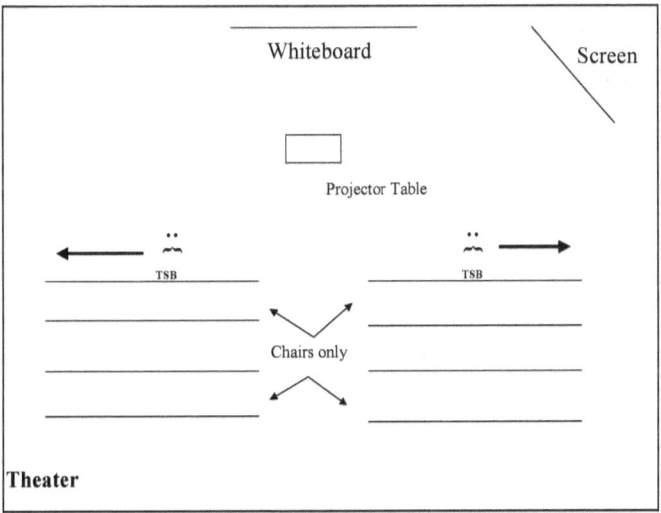

Troublesome buyers are indicated on the diagram as previous examples. Troublesome buyers (TSBs) can be anywhere on the front row. The farther they are from the center aisle, the better. It's much easier to ignore them, especially if you position yourself toward the center of the room. If they end up on the aisle, move into their personal space. In a theater style setup, the TSB won't even have the protection of a table, so you can easily move very close and increase his/her discomfort level.

Potential supporters should be placed evenly around the room and next to troublemakers. What you are looking for here is to broaden your impact by taking advantage of friendly buyers. They can help you with troublemakers, but more importantly they can reinforce what you are saying. Try not to let friendly buyers sit in a group. You don't want to create an "us against them" attitude within your buyer group. Spread your friendly buyers out to get the most benefit from them.

Flow of Conversation

Controlling the flow of the conversation during your presentation is YOUR job. During the early phases of your presentation, try to get as many people involved as possible. Speak directly to everyone in the audience if your group is 15 people or less. In larger groups, speak to about the same number of people and people that are dispersed throughout your audience. The next best thing to talking to someone is talking to the person next to him/her. You want everyone to feel a part of the presentation. Try to have the conversation move from one side of the room to the other.

Keep the conversation moving and do not let one person dominate. You can do this by asking open-ended questions (how, what, why, when) followed by a quick closed ended question (yes or no answer). Look directly at each person as they speak. Nod in agreement or acknowledge that you understand what they are saying. Step toward each person as they speak if you can. If you speak for more than 5 minutes without any interaction, you will probably lose your audience. Remember, they are going to remember more of what they say than what you say in the long run.

Taking Charge

There will be times when you will have to just take charge. The best way to explain this is by examples. I have provided some situations followed by proven solutions.

Situation:

Two people are having a conversation during your presentation

Solution:

You say, "Excuse me. Did one of you have a question?" If the answer is no, ask "Is there something that you would like to share with the group?" The answer will usually be "no". Then say, "May I continue?" If the answer is "yes", ask one of them by name to share. This can be done by just saying their name. Unless you have an incredibly rude group of buyers, you should not have this problem again. You have also established with the group that you expect and will demand appropriate behavior from the group in a polite and professional way.

Situation:

More than one person is talking at the same time.

Solution:

You should say, "Excuse me, but I would really like to hear what each of you has to say. Jane, why don't you go first and then, we'll hear from Bob." Be sure to assign a sequence or they will all start talking again.

Situation:

An internal issue that will not be solved with your solution such as policy, organization, or technology is dominating the conversation

Solution:

You should state the issue and the fact that this does need to be resolved. Then state that this is probably not the forum to resolve it. Perhaps they should schedule a meeting following this one to discuss the issue. You should assert that your solution does not address this internal issue or that your solution will work regardless of what they decide. You need to regain control of the audience and refocus them on your agenda.

Situation:

Someone walks in late to the meeting.

Solution:

Stop your presentation. Introduce yourself and any other people from your company. Find out who they are and let them get seated. When they are seated, briefly bring them up-to-date on where you are in your presentation. Offer to answer any questions during the break. Start up your presentation again.

Situation:

Someone keeps asking you to jump ahead in your presentation.

Solution:

You should tell him/her approximately how long it will be before you discuss their requested topic. Tell him/her again the sequence of your talk. You should have done this once at the beginning of your presentation. If they are not satisfied with this, suggest, "Perhaps you would like to leave and come back in 20 minutes (time before topic) when I will be covering that topic." You have now given him/her an out and you have also politely told him/her that you are not going to change the sequence of your presentation. In most cases he/she will not leave, but he/she will be quiet. At the introduction, you should have gone through the agenda and gotten agreement from the audience. However, not everyone will have paid attention.

Situation:

Someone is using offensive language.

Solution:

This has only happened to me a few times and I have to say it was quite unexpected. There is no excuse for this type of behavior from a buyer and you need to put a stop to it the first time it happens. You are perfectly within your right to ask them to stop. You should tell the buyer that you find their language offensive and that you would appreciate it if they would stop. If they don't stop, ask them to leave. You can bet other people were offended or embarrassed too.

Situation:

Handling objections to your product or service.

Solution:

Your marketing department should provide you with tactical statements to address the known objections to your product. Some will be the results of their competitive analysis. The number one rule is, don't get defensive! Your competitors planting objections about your product with your buyers is a common practice and you will be doing it too about their products. These objections will also alert you as to who you are competing against. Some of these objections will relate to known weaknesses in your product/service and are the result of one or more of your competitor's tactical strategies. Others may be the result of a competitor's strategic positioning that is not consistent with your company's direction. Be sure that you know the difference. Your marketing department should be providing competitive information and strategies for you to use. If you get caught with a new objection, you might try one of the following suggestions:

1. Try to think of a positive reason for why your product handles the "objection" the way it does.

2. Say something to the effect that as a customer, they will be able to provide input to the future of the product.

3. Express understanding of their objection and state that you will pass that suggestion along to your product people. Then quickly move on to the next topic.

The most important thing to remember when dealing with objections, do not dwell on them. Get past them as quickly as you can!

Analyzing your Audience

Only experience will make you adept at analyzing your audience. However, there are some signs that you can look for that will tell you where you stand with your buyers. Your buyers will be sending you signals without realizing it. Therefore, it is important to pay close attention to your buyers so that you can pick up on these signals. They are usually in the form of caution, positive or negative signals.

Caution Signals

Caution signals will indicate to you that things may not be what they seem. For example, the decision makers are not taking notes. This usually means that 1) they have confidence in their subordinates or 2) a vendor selection has already been made and you are there to provide "matrix fodder". In this case, watch the influencers to see if they are taking notes. If they are not, you have a serious problem. Either a decision has already been made or one is not going to be. If the influencers are taking notes, listen closely to their questions. Are they relevant questions? Do they relate to the buyer's value chain? If they do not, you have a problem. If this is true, the sales representative will need to ask some very direct questions during his/her follow-up to your presentation.

Another caution signal is the people are just too friendly. In some cases, the buyers may be very friendly people. On the other hand, their too friendly attitude might be a reflection of their guilt. If they have already made a decision, some buyers may feel uncomfortable with the situation and overreact in the friendly department. You can waste a lot of you time and your company's resources if you let buyers mislead you during an evaluation.

If only the lower level people are asking all of the questions, this is another caution signal. It shows a lack of interest on the part of the decision makers and the influencers. If you can't draw the decision makers into the presentation with open-ended questions, this is probably not a deal that you are going to win.

Positive signals

Positive signals will indicate to you that the buyers are interested in your product or service. Positive signals include: decision makers and influencers asking questions, notes are being taken, people are leaning forward in their chairs, people are nodding in agreement to a benefit, there is a high level of meaningful interaction, value-chain related questions are being asked and buyer's are repeating back your solution(s) to their problems(s).

Negative signals

Negative signals obviously alert you that you have a problem. If you are not getting any of the above-mentioned positive signals, that in itself is a negative signal. If buyers are not talking about specifics related to your product during a break, this is a negative signal. It could possibly be that you are just out of sync with your audience as discussed in Chapter 7 or it could mean there is no interest in your product.

Your buyer's body language will also send out negative signals. Buyers sitting back with their arms crossed over their chests, tells you that they are placing a barrier between you and them. If they are expressionless, they are probably not even listening to you. Your first reaction should be to liven up your presentation with a funny analogy or customer story for example. Get their attention and then start asking open-ended questions.

If the suggestions in Chapter 7 for the out sync problem do not work and nothing else seems to work, it usually best to take a break and talk one-on-one with a decision maker to find out what the problem is. It is not in your best interest to keep plugging along. You need to find out if there is something you can do. In a team-selling environment, the sales representative should lead the discussion with the decision maker. The product person should be included, but should initially listen rather than lead. The sales person should be managing the account..

Unfortunately, the problem may be you, the speaker. There may be times when your personality does not fit with the buyer group or your knowledge level is intimidating (another reason to avoid telling all). It has happened to me and to many of my peers when I was in sales support. It can happen to you. If a decision maker will not make any recommendations, he/she is basically telling you that he/she is not interested enough to be helpful. You should wrap up your presentation as quickly as possible. Be merciful. End it! Don't waste any more of your time and money on this buyer!

Summary

By controlling and analyzing your audience, you are making the best use of the limited time you have in front of a buyer. When this is done well, you can be confident that you have made this time beneficial to your buyer as well as to your company. Depending on the length of your sales cycle, you may only have one opportunity to sell your product. You want to make sure that the objectives you set prior to your presentation have been met. You do this by staying in control of your presentation and avoiding the pitfalls that can undermine your effectiveness prior to and during your talk.

Chapter 9

Seminars

Seminars offer a unique opportunity to reach several buyers with one presentation. It should not be considered to be a presentation that will automatically generate orders. In almost every situation, sales activities must follow in order to get a signed contract. A seminar should be treated as an opportunity to generate qualified interest in your company and your product or service. As a result, seminars and seminar presentations should be well thought out and planned to perfection. In many cases, this will be the first opportunity for your potential prospects or suspects to be exposed to your company. This chapter is not meant to be a guide to seminars, but it will cover some logistical issues as well as presentation techniques.

Logistics

You will want your seminar to flow very smoothly and require minimal effort from the audience. Unlike a specific buyer presentation, seminars are to pique interest not to make a buying decision. You want the experience to be interesting and fun for your audience.

You must have an agenda and be sure to follow it. It's even more important in a seminar setting to stick exactly to the time frames identified in the agenda. Always start exactly on time. Holding up the seminar for stragglers is very aggravating to those people who cared enough to be on time. Schedule frequent breaks, at least one per 75 minutes.. Keep breaks short unless your audience is moving from one room to another. Coffee breaks should only be 10 minutes. You do not want to lose momentum during a break

If possible, select a room setup with tables. It will be much more comfortable for your audience. Provide pads, pencils and water. Unless you are using your marketing materials as part of your presentation, pass out only the agenda, company information and a nametag when your guests arrive. Marketing materials can be placed on a table for people to take with them at the end of the presentation. You want people listening to you, not reading your marketing materials. This will also reduce your costs for seminars. There is no point in giving materials to someone who will minutes later throw them away. Interested people will take what they want.

Be sure that the welcome at the beginning of the seminar includes directions to the restrooms and the telephone. Reinforce that breaks will be short. Ask the audience to please turn off their beepers and cell phones.

113

Be sure that an attendee list is provided to each of the speakers. They will need to know the composition of their audience. The list should include name, title and company name. If at all possible, customers need to be identified and flagged as positive or negative.

Presentation Techniques

You have two objectives for your seminar presentation. First, you want to introduce your product/service to the audience. Second, you want to entertain them.

Your presentation should be very high level and very benefit oriented. Don't try to explain the "how's" and "why's" in a seminar. You will have an audience that is a mixture of application and technical, management and non-management and each will have their own set of specific needs. You want to concentrate on the features and benefits that will appeal to and be understood by a broad audience. The more customers illustrations that you can use, the more your audience will relate to your product and the more confident they will feel about it. You do want to involve your audience, but not to the degree that you would in a buyer situation. If someone asks a very detailed question that requires a very detailed answer, postpone answering it until a break. If the technology supporting your product can be separated from the application usage, have a separate session to discuss it. This will alert non-technical people that this session may be unsuitable for them to attend. There should be NO surprises to your audience in terms of the type of material covered in each of the sessions.

Probably the hardest part of doing a seminar is making it fun for the audience. You can use gimmicks such as having everyone introduce themselves to their neighbors to help break the ice. Don't go around the room and have everyone introduce himself or herself. Boring!! Entertaining analogs or fun visuals that relate and can invoke laughter are good. Don't feel that you have to be a comedian. If you are, that's great.

Summary

Seminars are a good way to introduce your company and product/service to a large number of potential buyers with one presentation. Just keep in mind that you cannot satisfy individual buyers needs with a seminar unless your product has a very narrow scope. Seminars need to be very professionally done and if they are to be a success, they should be entertaining.

Chapter 10

Follow-up

Follow-up correspondence is your way of reinforcing each presentation or meeting during the sales cycle. If you are like most people, you hate paperwork. This final chapter provides you with some guidelines for making this process less painful.

Written correspondence should follow each major event during the sales cycle. These events include your survey, your initial presentation, any follow-up presentations, any meetings or lengthy telephone conversations. With the exception of your first letter following your survey, each letter and accompanying document should be selling your product and reinforcing the benefits of your product to the buyer.

In the team-selling environment, both the sales representative and the product person should follow each event they are involved in with correspondence. These letters should be coordinated so that the major benefits are consistent. This chapter deals with the product person's letters only.

116

The easiest way to handle mass correspondence is to have predefined letters that can quickly be customized per buyers. With today's spreadsheets and word processing packages, customized letters can easily be produced with a mail merge. You should compose a letter for each type of situation that you usually encounter during your particular sales cycle. For example, you could have a survey letter, presentation letter, question and answer meeting letter, second presentation letter, etc.

Survey letter example"

Dear ():

I appreciated the opportunity to meet with you on (date). Learning about your (specific needs or company) will be extremely helpful to me in preparing for my presentation of (your product name) to (buyer's company name).

We are looking forward to presenting our solution to you. We believe in our product and upon seeing it, we think you will too.

Sincerely,

Name
Title
cc: (sales representative name)

This letter should not be a sales letter, but a thank you. It presents you as a polite professional and leaves a good impression on the buyer. This type of letter should be sent to each person that you interviewed. Use this letter as a guide. Your letter should sound like you! However, keep it brief.

Presentation 1 Letter Example

Dear ():

It was a pleasure seeing you again during our presentation on (your product's name). Based upon our earlier survey, plus our presentation and discussion, we feel confident that our product is the best solution for your needs.

I have attached a summary of the major points that we agreed would be of benefit to (buyer's company name). (Also, identify any specific follow-up that the buyer requested.)

Thank for your participation (last week/this week). It added so much to my presentation. As your evaluation progresses, please do not hesitate to call or email if I may be of help.

Sincerely,

Name
Title
cc: (sales representative)

Your summary of major points should be a standard list that highlights the major benefits of your product. Bullet format works best. Each bullet should contain a feature and a benefit. In addition to your standard list, you should add some specific features and benefits using their terminology to customize the list. This is also an excellent way to get the competitive advantages that you talked about during your presentation into print.

Being timely with your follow-up correspondence is the only way it will be meaningful to the buyer and make him/her feel that he/she is special.

Create letters and worksheets for each situation that occurs during your typical sales cycle. Each letter following your survey letter should have a feature/benefit orientation. If you have a question and answer session, you should include these in an attachment. Do not expect the buyer to make notes that are complete or accurate. Your follow-up correspondence will ensure that your buyers have all the facts they need to make a decision in your favor.

You do not need to produce a unique letter for each attendee. The same letter can be used. You may want to develop an executive level series of letters and a non-executive series. Any further breakdown will just be time consuming with little added benefit. The main idea is to get the letters out to everyone who was involved in the event. It is also a good idea to send a thank you email to the administrative person that helped you with the logistics. Email is a perfectly accepted form of correspondence. You do need to be sure that your company's office technology is compatible with your buyer's. Otherwise you will need to send text messages that aren't that attractive or readable.

Summary

Follow-up correspondence is essential to presenting a totally professional image of you and your company. It should be done following each major event to each participant and as soon after the event as possible.

www.ingramcontent.com/pod-product-compliance
Lightning Source LLC
Chambersburg PA
CBHW022005170526
45157CB00003B/1151